The OECD, Globalisation and Education Policy

ISSUES IN HIGHER EDUCATION
Series Editor: GUY NEAVE, International Association of Universities, Paris, France

Editorial Advisory Board:

Other titles in the series include

The OECD, Globalisation and Education Policy

Miriam Henry
Queensland University of Technology, Red Hill, Australia

Bob Lingard
University of Queensland, Brisbane, Australia

Fazal Rizvi
Monash University, Clayton, Australia

Sandra Taylor
Queensland University of Technology, Red Hill, Australia

IAU
PRESS

Published for the IAU Press
PERGAMON

An imprint of Elsevier Science
Amsterdam – London – New York – Oxford – Paris – Shannon – Tokyo

ELSEVIER SCIENCE Ltd
The Boulevard, Langford Lane
Kidlington, Oxford OX5 1GB, UK

First edition 2001

Library of Congress Cataloging in Publication Data
A catalog record from the Library of Congress has been applied for.

British Library Cataloguing in Publication Data
A catalogue record has been applied for.

ISBN: 0 08 043449 5

⊗ The paper used in this publication meets the requirements of ANSI/NISO Z39.48-1992 (Permanence of Paper).
Printed in The Netherlands.

The IAU

The International Association of Universities (IAU), founded in 1950, is a worldwide organization with member institutions in over 120 countries, which cooperates with a vast network of international, regional and national bodies. Its permanent Secretariat, the International Universities Bureau, is located at UNESCO, Paris, and provides a wide variety of services to Member Institutions and. to the international higher education community at large.

Activities and Services

* IAU-UNESCO Information Centre on Higher Education
* International Information Networks
* Meetings and seminars
* Research and studies
* Promotion of academic mobility and cooperation
* Credential evaluation
* Consultancy
* Exchange of publications and materials

Publications

* International Handbook of Universities
* World List of Universities
* Issues in Higher Education (monographs)
* Papers and Reports
* Higher Education Policy (quarterly)
* IAU Bulletin (bimonthly)

Introduction to Issues in Higher Education

For the past quarter century, higher education has been high on the agenda of governments and central to the fortune of nations. Similarly, this same period has seen quite massive changes in direction, in the complexity of systems, in the underlying rationale which has accompanied such changes and in the sheer size of the enterprise in terms of students, staff and budgets, not to mention social and economic purpose. It is not surprising then that the study of higher education itself has broadened and now encompasses some 20 different disciplines, ranging from Anthropology through to Women's Studies, catch with its own particular paradigms, methodologies and perspectives.

Against this background, the comparative analysis of higher education policy which has always occupied a crucial place in understanding the contextual setting of reform in individual countries, has acquired a new significance as the pace of "internationalization" itself quickens. There are many reasons why this should be so: the creation of new economic blocs and, in the case of Europe, the gradual emergence of a transnational policy for higher education across the EC countries; the triumph of one industrial ethnic and the collapse of another, the rise of new economies in Asia, etc. The breakdown of a seemingly established order has ushered in a renewed interest in other models of higher education and in how other nations are going about tackling often similar issues and testing theories in the field of higher education policy which are of current and practical concern to its main constituencies — national and institutional leadership, administrators, teachers, those researching in this domain and students. As a series, it will focus on both advanced industrial and also on developing systems of higher education.

Issues in Higher Education will be resolutely comparative in its approach and will actively encourage original studies which are firmly based around an international perspective. Individual volumes will be based on a minimum of two different countries so as to bring out the variations occurring in a given problématique. Every encouragement will be given to the drawing of clear and explicit comparisons between the higher education systems covered.

As the editor, I wish to thank the members of the Educational Advisory Board for their part in developing this series. They are:

Jose Joaquim Brunner, *FLACSO (Latin American Faculty for Social Sciences), Santiago, Chile*

Burton R. Clark, *Emeritus Professor, Graduate School of Education, University of California, Los Angeles*

Dan Levy, *Professor of Educational Administration and Policy Studies, State University of New York, Albany, USA*

Lynn Meek, *Department of Public Administration and Studies in Higher Education, University of New England Armidale, New South Wales, Australia*

Hassan Mekouar, *University Mohammed II, Morocco*

Keto Mshigeni, *The Graduate School, University of Dar-es-Salaam, Tanzania*

Agilakpa Sawyerr, *African Association of Universities, Accra, Ghana*

Ulrich Teichler, *Director of the Research Centre for Higher Education and the Labour Market, University of Kassel, Germany*

Morikazu Ushiogi, *Department of Higher Education, Nagoya University, Japan*

Frans van Vught, *Center for Higher Education Policy Studies, University of Twente, Enschede, The Netherlands*

Fang Min Wei, *Institute of Higher Education at Beijing University, The People's Republic of China*

GUY NEAVE
International Association of Universities
Paris, France

Contents

Acknowledgements

The research project on which this book is based was funded by the Australian Research Council from 1995 to 1997 and we are grateful for their support.

We would like to thank all those people in Paris, London and Australia who gave so generously of their time to speak with us — sometimes on more than one occasion. In particular we would like to thank Alan Wagner at the OECD who "paved the way" for us, and also Emma Forbes who provided us with many OECD reports and documents on which we have based our analysis.

We would also like to thank the participants who attended the symposium on "The OECD, globalisation and education policy making" — held in Brisbane in June 1997. The papers presented and discussion at that symposium, as well as with colleagues near and far, have helped to shape our ideas.

Writing the book from three institutional locations in two cities has not been an easy task and we are grateful for the support of colleagues, family and friends. Finally we would particularly like to thank Merle Warry and Joy Doherty for their administrative and research assistance, Meredith Sadler for her care with editing and formatting of the manuscript, Nicholas Lingard for getting the text ready for the publisher, and Guy Neave for his punctillious editorial polishing.

1

Why the OECD?

"The OECD is not a supranational organisation but a place where policy makers can meet and discuss their problems, where governments can compare their points of view and experience. The secretariat is there to find and point out the way to go, to act as a catalyst. Its role is not academic; nor does it have the authority to impose its ideas. Its power lies in its capacity for intellectual persuasion."

(OECD, 1985a, p. 3)

"The Organisation for Economic Co-operation and Development, based in Paris, France, is a unique forum permitting governments of the industrialised democracies to study and formulate the best policies possible in all economic and social spheres."

(OECD, 1994a, p. 7)

Why, one might ask, a book on globalisation and education policy which pivots around this rather stolid-sounding, economic-oriented international organisation? There are several answers, for behind the bland self-descriptions above are a number of complex issues about how an international organisation such as the OECD 'works', comes to formulate 'best policies possible' and relates to the policy-making machinery of national governments. These issues will become clarified over subsequent chapters as we attempt to analyse what such a description may mean in terms of the OECD's educational agenda and policy-making processes. This task, however, is linked to a broader analytical intent: to understand the changing nature of policy making in the context of the epochal shift now most often described as the era of globalisation. Such an analysis is at the core of this book. The purpose of this first chapter, then, is to set the scene for the book: to say something about how it came to be written;

to describe its theoretical pedigree; to delineate some of the methodological problems we have encountered along the way; and to say something about the OECD itself.

Background and Theoretical Framework

This book had its genesis in a three-year research study funded by the Australian Research Council (ARC) looking at the role of the OECD in shaping education policy directions in Australia since 1984 in the context of globalisation. We had become interested in the OECD because in a prior study on policy changes in Australian higher education and schooling the Organisation seemed to be a significant reference point for a number of respondents. Australia's comparative position within the OECD league table on various policy issues was, for example, often mentioned. Others referred to the importance of the OECD in initiating a new policy paradigm — micro-economically framed version of human capital theory — as the justification for moves towards a *national* policy focus in higher education and schooling in a country where states and territories jealously guarded their educational terrain. Yet others alerted us to the emergence of an embryonic global policy community, in which the OECD was a significant player, impacting upon policy making in education within Australia.

Clearly, the OECD's role was contingent rather than direct, yet there seemed to be little research about how the Organisation exerted influence. In our ARC research, we wished to probe the 'mechanisms of persuasion' through a case study of the relationship between the OECD and educational policy directions in Australia. Australia was selected in part because of our own location as researchers, but also because Australia appears to have been a particularly active player within the OECD educational arena. At the same time, we wished to raise broader questions about the nature and scope of the relationship between international agencies and educational policy development at the level of the nation-state — questions which in turn involved a critical interrogation of the emerging literature on globalisation.

When we formulated the OECD study, globalisation was not the fashionable and ubiquitous research topic it has subsequently become, though globalisation as an idea has existed in sociological theory for most of this century. More recently, the idea of globalisation has been used to describe relentless calls for change in social and political systems and economic structures. The ensuing developments have variously been referred to as a move toward a global market economy, the new internationalism, the internationalisation of labour, and the emergence of a global culture and transnational or supranational policy structures. These developments have given rise to a range of complex issues relating to the future of nation states and their capacity to control their own policy destinies.

There is in all of this a degree of terminological and conceptual confusion, so we will indicate our own use of terms in this book, though without entering the debates about the meaning of globalisation and without preempting the elaboration of some of the theoretical issues surrounding globalisation in Chapter 2. Our view is that globalisation has provided the impetus for a new regionalism (for example, the North American Free Trade Agreement or the Asia–Pacific Economic Cooperation) and new supranational organisations (for example, the institutions of the European Union such as the European Parliament or the European Court of Justice whose constituency is not tied to national boundaries). Globalisation has also acted to destabilise the nation-state, making its boundaries more porous, and to change the way international organisations operate. Thus the OECD, for example, once depicting itself as "not a supranational organisation", now describes itself as a "global intergovernmental" economic organisation (OECD, 1998a, p. 5) in recognition of a changing context. We use the term 'supranational' or 'global' — rather than transnational or international — to describe the linkages between these reconstituted units even though the relationship may still be between nation-states or between units which are not supranational in terms of political organisation.

In education, we have witnessed the emergence of global flows of educational ideas around issues of educational governance and purposes. But how these ideas become embedded within the policy agendas of nation states is neither well researched nor well described. One assumption underlying our study, however, was that the OECD, as a 'globalising agency', may play an important role in this process. The study, then, attempted to bring together three contexts of policy making: the OECD as an international organisation with some influence in education; the context of globalisation; and the context of national policy making with specific reference to Australia. There has been little research on the OECD itself, which is paradoxical given the Organisation's prolific research output. We wanted to investigate how the OECD exerted influence in education policy and how the context of globalisation might be impacting on both national and international levels of policy making.

This book reports on the findings of that study and in so doing opens up new ground. For even now, little has been done on the effects of globalisation on educational policy, and even less on the specific workings of globalisation through international organisations in education. Spring (1998) paid some attention to the role of the OECD and World Bank in globalising and legitimating the conception of education as a producer of human capital for the global economy, thereby confirming the often hailed view of the Organisation as a neo-liberal demon. Scott (1998) examined the ways in which higher education may be implicated as "creator, interpreter and sufferer" of globalisation, though much of the book in fact deals with the internationalisation of higher education — a somewhat different phenomenon as we have argued elsewhere (Taylor et al., 1997, ch. 4) — particularly in the context of the European Union. Deacon et al. (1997) have begun to pursue and theorise the relationships between globalisation and social policy within nations, again with particular reference to the European Union, arguing that national social policies are increasingly being affected by policy frameworks articulated and disseminated by supranational political units.

They observe, however, that the linkages between globalisation and social policy within nations remain "an under-theorised and under-researched topic within the subject of social policy" (p. 1). We would reiterate their observation in respect of educational policy. This book represents a contribution to that endeavour.

Our assumptions are as follows. The OECD has been a key articulator of a predominantly neo-liberal reading of globalisation. But it is a complex organisation, and its work and influence in the field of education reflect more ambiguous stances which may contribute to both strengthening and to undermining national policy making. At the same time, notions of national policy making and indeed national interest — always problematic given the historical specificity and fragility of the nation state and the competing interests within it — have become increasingly fraught given the simultaneous impulses of globalisation towards political and economic concentration and cultural homogenisation on the one hand, and on the other hand new forms of regionalism, localism, diasporic connections and cultural heterogeneity. At a normative level, we would support the contention that globalisation constitutes an emergent form of Western imperialism carried through cultural institutions such as education. We note Waters' point that globalisation does not infer that the entire globe has become Westernised, but rather that all spheres of social life must establish their position "in relation to the capitalist West" (Waters, 1995, p. 3). Western capitalism, then, has become the reference point against which nation-states establish their policy options. This is demonstrated in some of the case studies in this book, and due care has been taken not to oversimplify the range of positions on the neo-liberal/social-democratic spectrum. Given that globalisation works ideologically as well as materially, its normative presuppositions may be — and in our view ought to be — challenged. A fundamental political task, then, is to establish parameters for education policy making 'in the national interest' which recognise the force of global change while challenging the ideological determinism of globalisation rhetoric, that is, the neo-liberal reading, as the only option.

Our analysis is located within the field of education policy sociology, described by Ozga (1987, p. 144) as "rooted in the social science tradition, historically informed and drawing on qualitative and illuminative techniques". Policy sociology — or critical policy analysis as it is referred to in Australia and the US is a multidisciplinary field of inquiry which brings the critical and structural insights of sociology to more traditional approaches in education policy analysis. In contrast to more technical and linear views of policy found particularly in the educational administration literature and in many bureaucratic approaches to policy making, this approach views as problematic those very processes which more conventional accounts take for granted. In particular, the frame of reference of policy sociology draws attention to the sources and distribution of power. It raises questions about the shaping and timing of policy agendas in ways which are particularly pertinent to this study given our theoretical concern with understanding, not simply 'the facts' about policy developments and processes, but also under what circumstances and with what effect such developments have assumed their particular shape.

Given this framework, the research techniques employed were necessarily qualitative in nature. Interview notes and transcripts, policy documents, reports and memoranda both published and unpublished, constituted the textual base for analysis. Surprisingly, the OECD has no educational library as such — or at least not at the time we conducted our study. However, draft programmes of work and back issues of newsletters for the period we were interested in were made available and are drawn upon extensively at times. Here it is also appropriate to acknowledge a particular debt to George Papadopoulos's (Papadopoulos, 1994) authoritative historical account of education in the OECD which we also draw upon extensively at times. Papadopoulos was the founding deputy director for education within the Organisation and his account has been invaluable in providing the historical context for our contemporary study.

Loosely structured interviews were conducted with key players from key locations, that is within the OECD and other international organisations in Paris and, given the Australian location, within the Australian policy making and research communities. Additionally, some interviews were conducted with UK researchers to provide some points of comparison. Some of our interviews were formally recorded and transcribed; others took place informally, with notes written up more extensively as soon as possible afterwards. How to designate these interviews in this book has been problematic. Some of our interviewees, either by virtue of their position or perhaps by their personality, were unconcerned about going 'on record'. Some so carefully hid behind their position as to reveal little but the public face of their work and perhaps would have had little concern about being publicly associated with it. Others, however, gave us more critical insights, confident that their views could be expressed anonymously. For consistency then, all interviews have been coded — designated in the text by bracketed numbers, with the letters IT and IN denoting interview transcriptions and notes respectively. The interviews have been classified into four broad groups: OECD secretariat; other international organisations; bureaucrats/government policy analysts; consultants. These labels accompany the bracketed numbers in the text. Something is lost in this classification process of course; sometimes it might be useful to know whether an observation was made by a deputy director, an 'ordinary' secretariat member, a current or a past consultant, a male or a female and so on. Ensuring anonymity was our key concern. Interviews were conducted over the three-year period 1995–1997; the year noted in interview reference, then, denotes the year the interview was conducted.

Worth noting perhaps are some of the methodological problems we encountered in the course of the study, and some of the gaps. The OECD is a bewilderingly complicated organisation, connecting to both government and research communities and producing a vast number of publications, reports, reviews and other documents. The problems of selection were considerable. Educational work in the OECD is organised around four 'programmes' within the

Directorate of Education, Employment, Labour and Social Affairs (DEELSA). The four educational programmes relate to education and training policy, educational research and innovation, institutional management in higher education, and educational buildings. To keep the study manageable, we restricted our interviews to members of the education side of the Directorate only, and omitted from our focus the programme on educational buildings. And while our interviews, 68 in all, spanned a reasonable range of OECD secretariat members, Australian government (or former government) bureaucrats and policy makers, and OECD consultants/researchers both in Australia and the UK, representatives from two important groups are missing: the OECD's two statutory consultative committees, for Business and Industry and the Trade Unions respectively. We can only plead shortage of time and energy and hope that the gap does not constitute a serious flaw in our analysis.

There are also gaps relating to factors of culture and gender which we will only touch on here. Several of our interviewees commented on the Anglophone character of the OECD, despite its two official languages of French and English and despite the prominent role of the Nordic countries, at least within the education section. With respect to gender, while our research team was gender balanced, our interview pool was not: of the 68 interviews conducted, 16 were with women. Among the pool of consultants we had difficulty finding women.

Most vexing was the selection of themes to study given the range of the OECD's educational work. Our time-frame provided one selective mechanism, and we have already mentioned the omission of work relating to the educational buildings programme. Another selective mechanism, at least in terms of this book, is its publication in a series devoted to the study of higher education including, in this era of globalisation, vocational education and training. As a consequence, issues relating to schooling and early childhood education to a large extent have had to be neglected here. Beyond that, the themes which captured our attention and which have formed the focal points for the chapters of this book were those which seemed to reveal various facets of globalisation processes working in and through education.

Organisation of the Book

The groundwork for the book is set out in the following three chapters. Chapter 2 explores the terrain of globalisation theory and the implications of globalisation, both in its material and ideological manifestations, for educational policy making and governance. The emphasis is on political globalisation given our concern with policy priorities and processes. Chapter 3, attempts to link three arenas: globalisation; educational policy making; and the role of the OECD as an international organisation. Chapter 4, in examining ideological tensions in the OECD's educational work, explores the changing ways economic and social agendas play off against each other in the face of globalisation pressures. The case studies in Chapters 5, 6 and 7 relate to educational indicators, vocational education and training, and developments in higher education respectively. Chapters 6 and 7 are largely but not exclusively based on Australian policy developments and links between Australian and OECD policy arenas. But they are also case studies of the globalisation of education policy and have something to say about the changing nature of policy making over and beyond the Australian experience. The final chapter, on the future of the OECD, poses more normative questions about the locus of education policy making in a new global — as distinct from a new international — order.

So now to the OECD itself, and to the question of how education fits within this essentially economically oriented Organisation. [1] The next section, then, begins to 'unravel' the OECD in terms of its structure and mode of operating, though more is said about these issues in Chapter 3. The description which follows is somewhat abbreviated, aimed simply at sketching out the Organisation's most salient features in order to ground subsequent discussion and to flag some of the more complex issues which are pursued in greater depth later.

Unravelling the OECD

What is the OECD? It is, simultaneously, a geographic entity, an organisational structure, a policy-making forum, a network of policy makers, researchers and consultants, and a sphere of influence. Salubriously located in the Chateau de la Muette and surrounding buildings in the upmarket sixteenth arrondissement of Paris, it comprises representatives from governments of 29 mostly wealthy Western countries though, since the mid 1990s, the membership has changed. Hence the founding members of the OECD in 1961 were Austria, Belgium, Canada, Denmark, France, Germany, Greece, Iceland, Ireland, Italy, Luxembourg, the Netherlands, Norway, Portugal, Spain, Sweden, Switzerland, Turkey, the United Kingdom and the United States. Over the next decade or so, these were joined by Japan (1964), Finland (1969), Australia (1971) and New Zealand (1973). A twenty-year gap separates these essentially European or Anglophone members from the new arrivals — Mexico (1994), the Czech Republic (1995) and then, in 1996, Hungary, Korea and Poland. More will be said about the implications of this changing membership later, simply noting here that current descriptors still convey a sense of privilege: "a 'think tank' to some, a 'Rich Man's Club' to others" (Istance, 1996, p. 91); "a kind of international think-tank ... an English-speaking, northern European organisation" (IN 16, secretariat, 1995); "a policy-makers policy making unit" (IN 30, consultant, 1996); "a European club, not easily changed" (IN 44, bureaucrat, 1996). The Organisation describes itself as:

---

[1] The distinction between the OECD's general and educational aspects is not always clear or relevant throughout the book. However, it should be borne in mind that the central focus of the book is on *education* policy making; for convenience, we often refer simply to 'the OECD' when in fact a more a precise allusion to its educational work is intended.

"a club of like-minded countries. It is rich, in that OECD countries produce two thirds of the world's goods and services, but it is not an exclusive club. Essentially, membership is limited only by a country's commitment to a market economy and a pluralistic democracy."

(OECD undated, circa 1997, p. 4)

The OECD was established in 1961 evolving out of the Organisation for European Economic Cooperation (OEEC) funded under the Marshall Plan for the economic reconstruction of Europe. While the OEEC was a European organisation, it remained essentially a US initiative. And although membership of the newly formed OECD embraced North America, the move to Paris signified a more European orientation (Archer, 1994). Still, the shadow of the US — contributing 25% of the Organisation's budget — hovers over the OECD, particularly in the dominant 'tone' set by US versions of market liberalism. Additionally, the US makes key interventions from time to time: US pressure brought both Japan and Mexico into the 'club'; US intervention ensured the appointment of a North American (Canadian) Secretary-General, Donald Johnson, to replace the outgoing (French) Jean-Claude Paye in 1996; and in education, it was largely at the insistence of the US and against considerable internal opposition that the controversial project on educational indicators was initiated. Haas's description (Haas, 1990, p. 159) of the founding of the OECD as "a rather incoherent compromise between the United States and the European members" remains valid, as does his observation that most international organisations have their own superpower "capable of playing a hegemonic role if it chose to do so" (p. 57).

The Organisation's ideological positioning is reflected in its formal requirement for membership, namely the commitment to a market economy and pluralistic democracy. Recently, 'respect for human rights' was added to membership requirements, a point we will return to in Chapter 4. This ideological stance sets it apart from, say, the UN system with its more heterogeneous membership of rich and poor nations, more volatile ideological mix and more internally contradictory policy concerns. Yet, the OECD is hardly a homogeneous club, displaying, for example, significant differences between Anglo-Saxon versions of economic liberalism and more tempered notions of the market held by many of its European and Scandinavian members. In educational terms, such ideological differences translate into a "Nordic/European–Anglo-Saxon cultural divide" (IN 2: secretariat, 1995) which ripples through into policy conflicts, for example around issues of educational selection (in crude terms, Anglo-Saxon meritocrats versus Scandinavian egalitarians) or multiculturalism (again crudely, the homogeneous European countries versus the culturally more diverse Anglo-Saxon countries). Nevertheless, an overarching homogeneity is evident in terms of the Organisation's fundamental principles, its membership, its consensus-based decision making processes, and style. Thus overt conflict is generally converted into 'robust' discussion, behind closed doors, oiled by the machinery of high diplomatese and French *politesse* (IN 33: consultant, 1996).

This capacity to contain and retain members is important. Unlike many other international agencies, the OECD has no prescriptive mandate over its member countries. It operates through a process of "consensus building through 'peer pressure' ... [which] encourages Members to practice [sic] transparency, provide explanations and justifications for policy, and engage in critical self-appraisal" (OECD, 1998a, p. 102). Elsewhere, it has described itself as a place for reflection and discussion, research and analysis "that may often help governments shape policy", exerting influence through processes of "mutual examination by governments, multilateral surveillance and peer pressure to conform or reform" (OECD, undated, p. 10). How these processes of surveillance and peer pressure may work is a theme running throughout the book and is developed further in Chapter 3.

Structurally, the OECD works through an elaborate system of directorates, committees and boards, at the apex of which is a Council comprising representatives from each member country — normally at ambassadorial or ministerial levels. The European Commission is also represented on Council. OECD ambassadors — normally high ranking bureaucrats — and Ministers are generally from departments of Foreign Affairs or Finance. As a glance at the incumbents of these positions shows, this is indeed also a male, as well as Western, club. For example, in 1998 of the thirty OECD ambassadors (i.e. including the EC) only two were women and of the twenty-two secretariat heads, three were women (OECD, 1998a, pp. 110–112).

At first, no independent structural location for education existed within the Organisation, though there was always an "inferred role" (Papadopoulos, 1994, p. 11) deriving from early human capital formulations of links between economic productivity and educational investment, then somewhat narrowly conceived in terms of boosting scientific and technological personnel capacity and, by extension, of improving and expanding science and mathematics education in schools. Initially, education-related activities were carried out under the ambit of the Office for Scientific and Technical Personnel which in turn grew out of the former OEEC's pivotal work in mapping the technological gap between Europe and North America against the broader backdrop, in the post sputnik years of the cold war, of US disquiet about the Soviet Union's technological superiority (Papadopoulos, 1994; Ch. 2; Istance, 1996, p. 91).

The structural basis for education as an independent activity in the OECD, which persists to this day, came with the establishment in 1968 of the Centre for Educational Research and Innovation (CERI, initially with Ford Foundation and Shell funding), and the replacement of the Committee for Scientific and Technical Personnel with the Education Committee in 1970. These developments reflected a growing recognition within the Organisation of the qualitative aspects of economic growth "as an instrument for creating better conditions of life" (Papadopoulos, 1994, p. 64) together with a more comprehensive view of education's multiple purposes. By 1970, the Organisation had come to the view that "the full range of objectives of education had to be taken into account if the educational activities of the Organisation were to make their rightful contribution to economic policy" (*ibid.*, p. 64).

Describing the structural arrangements for education is, however, difficult — they are complicated and they provide only a rough map of how things work in practice. At the outset, CERI and the Education Committee were located within the Directorate for Scientific Affairs headed by Ron Gass, with George Papadopoulos as Deputy Director for Education. In 1975, education became part of a new Directorate for Social Affairs, Manpower and Education with Gass as Director. Gass, according to Papadopoulos (1994, p. 122), was one of the key figures in promoting a social agenda for the Organisation, reflected in the increased remit for social policy in the new Directorate. In 1991, that Directorate was renamed the Directorate for Education, Employment, Labour and Social Affairs (DEELSA), by which stage both Gass and Papadopoulos had retired, being replaced by Tom Alexander as Director and Malcolm Skilbeck as Deputy Director of Education.

According to Papadopoulos, the shift from the Science location occurred as on the one hand science policy moved away from the original concern with scientific and technical personnel, and on the other hand as unemployment pressures started to move education and training concerns closer towards the Organisation's main concerns with manpower and employment policies (p. 122). Papadopoulos argues that the new location gave a fresh impetus to education's social concerns, though not all agree with that view. Istance (1996) for example pointed out that while the new location consolidated a constituency of government education departments in member countries, the latter "exercised a largely conventional influence, curtailing the potential of the Centre for Educational Research and Innovation within the OECD to realise its title almost from the time of its inception" (p. 93). Istance's views are shared by some within the secretariat: "Education's placement in DEELSA was not wise. Its original location in Science and Technology meant the long term perspective was always in mind. Now, the short-term concerns with labour market and employment tend to dominate" (IN 2: secretariat, 1995). Others disagree: "DEELSA was created to effect better coherence of policies and a good cross-cutting department was needed" (IN 7: secretariat, 1995).

DEELSA itself comprises two major committees, the Education Committee and the Committee for Employment, Labour and Social Affairs. Education work is organised under four programmes: those of the Education Committee and the CERI Governing Board; and the more specialist programmes of Educational Buildings (PEB) and Institutional Management in Higher Education (IMHE), each with its own Steering Group. PEB is responsible to the Education Committee, while the other three programmes report directly to the Deputy Director for Education and ultimately to the Director of DEELSA, who is also the head of the CERI Governing Board.

Each of these programmes is funded in somewhat different ways, and links to member countries or institutions in different ways. The Education Committee is strictly governmental, while CERI Governing Board is partially non-governmental. PEB links to sub-national political units rather than national governments, while membership of IMHE is based on individual institutions, mostly universities.

There are two funding categories within the Organisation. Part I programmes are fully supported by core funding. Part II programmes may be partially or totally reliant on contributions from participating governments or institutions. Both PEB and IMHE are in the latter category. CERI draws mostly on core funding. Some of its activities, however, are subsidised by contributions from countries choosing to participate in those activities. In turn, governments may seek contributions to priority projects from their own constituencies. Thus, Australia's participation in the CERI project on performance indicators drew on contributions from Commonwealth and State governments, the Australian Vice Chancellors' Committee and the Schools and Vocational Education and Training Sectors. To some extent, then, CERI has more scope to do things differently and to do different things. However, all education programmes rely increasingly on in-kind contributions from member countries, in the form of contributed papers, participation at meetings, co-hosting meetings, and so forth. These activities are not funded out of the OECD budget, but rather from the budgets of authorities or other in-country sources. Education programmes do not have a permanent status within the Organisation. They are mandated by Council every five years. That the four programmes have survived is a testimony to their perceived legitimacy in the eyes of participating countries or institutions. If the governments did not approve of the work — accord it high enough priority, believe it was not relevant to their own needs, consider that the value added through the OECD was insufficient — the Council could stop all of the programmes including those not funded directly by governments.

The Education Committee has oversight of the Education and Training Division whose work, in theory, is essentially policy oriented. One of the significant responsibilities of this Division is the conduct of the reviews of country education systems and, more latterly, thematic reviews. We will say more about the significance of these reviews in subsequent chapters. CERI has more of a research focus. Initially established to consider curriculum issues, its work has gone on to encompass a very wide range of educational issues, one of the most recent being the development of the educational indicators project on which more will be said in Chapter 5. As with the Organisation more generally, the division of labour remains highly sex-based, with "men getting the important, economically oriented tasks, and women the softer ones such as the environment" (IN 40: consultant, 1996). 'Feminism is hardly an issue' (IN 5: secretariat, 1995). There is an EEO-designated position within DEELSA which, while reporting to the Director has no structural location within the Division, no separate budget and little influence on programmes and activities. Hence, it is claimed, "women's issues never get off the ground. All agree they are important, but never quite important enough to fund or sponsor, so things don't change there" (IN 44: bureaucrat, 1996).

In 1994 a Unit for Education Statistics and Indicators, the INES group, was formed drawing on resources from the Education Committee and CERI. Initially located within CERI, it recently became part of a separate Statistics and Indicators Division within DEELSA. The Statistics and Indicators Division attracts substantial additional resources from governments, particularly the US, to advance development work aimed at strengthening the indicators base. The work of this Division also connects with other, externally funded initiatives, for example the World Education Indicators, by which OECD's indicators framework is extended to ten developing, non-Member countries funded by the World Bank in co-operation with UNESCO, as well as work on Adult Literacy, funded by the US and Canada with local costs of survey administration covered by participating countries. CERI retains the function of carrying out the initial research for educational indicators, with INES being responsible for their development into operational indicators. However, this division of labour is confusing — to insiders and outsiders alike — as is the distinction between policy and research. In practice the divisions have less practical application than a formal organisational chart might suggest.

Thus, the Organisation's *modus operandi* belies its formal bureaucratic structure, in a number of significant ways. First, work is effectively organised around projects, themes, activities or programmes (the terminology varies) rather than structures: "teams are developed according to the problems which emerge" (IN 7: secretariat, 1995). Structural location is relatively unimportant. Secondly, the Organisation makes much of the principle of 'horizontality', designed to cut across formal structures. Horizontality poses the idea of cross-linkages between directorates, programmes and committees to facilitate multidisciplinary analyses of complex problems. Such horizontality emerged for example in the ambitious OECD Jobs Study on employment and unemployment which linked DEELSA's two committees with each other and with the Directorate for Financial, Fiscal and Enterprise Affairs. The involvement of CERI in an Education Committee conference on reforms to vocational education and training, is another example. Thirdly, while essentially an intergovernmental organisation, the OECD in some important ways takes on a more inchoate, non-governmental aspect. Its constituencies are not straightforwardly the *governments* of member countries. For instance, IMHE membership is institution-based; about one third of CERI Board members, though government nominated, are not government officials, which gives the Board "a somewhat different flavour from the Education Committee" (IT 56: secretariat, 1996), and much of CERI's work relates to research communities rather than governments. Secretariat members (*administrateurs*) combine functions of research, policy analysis and administration which involve them in a network of consultants and researchers independent of formal governmental channels. In principle then, this is a very fluid mode of operating. Indeed, according to Papadopoulos (1994, p. 15), it is precisely this flexibility, the "problem-oriented, *à la carte* approach to co-operation", and the professionalism of the small Secretariat with its wide-ranging linkages to national administrations, the research community, business and the unions, which provide the key to the Organisation's success.

Let us return now to education's niche in the OECD, and to questions of how educational work is produced, used and disseminated. Education has no permanent status within the Organisation. Rather, its existence is mandated by Council every five years under an organising theme — for example, "High Quality Education and Training for All'' for the period 1991–1996 (OECD, 1992e), and currently "Lifelong Learning For All" (OECD, 1996d). Entwined in these descriptors are two strands, an economic and a social, which have surfaced in different ways and at different times. In essence then, as Papadopoulos's (1994) account reveals, how education has been legitimised as an OECD activity is a story of economists being persuaded about the importance of investing in education in a context where economic prosperity and social enhancement were seen as parallel objectives.

Conversely, it is also a story about educationists embracing an essentially economic rationale while at the same time asserting an independent stance. This issue will be explored in detail in Chapter 4. Suffice to note that much more than a simple relationship between education and economic development has been staked out, which is evident from the sheer volume and breadth of programmes, projects, activities and publications produced by the education section within the OECD. Collectively, this represents a broader and more contested *corpus* than the OECD's economic mission might indicate. Included, to take just some examples, are reports and analyses of: school pedagogy and curriculum; the use of school buildings; educational disadvantage and advantage; multicultural education; girls and education; linguistic diversity; alternative education; school improvement and effectiveness; early childhood education; links between school and work; school–community relations; youth employment; youth at risk; people with disabilities; teacher education; educational and performance indicators; the economics of education; educational policy; educational technology; educational planning; educational management; mass higher education; vocational education and training; recurrent education and lifelong learning; adult literacy. In addition are a large number of reviews of the education systems in both member and non-member countries, as well as more recent thematic reviews on the first years of tertiary education and the transition from initial education into work which will be analysed further in Chapters 6 and 7.

How this work has been produced, disseminated and used, how countries use the forums of the OECD raises technical questions of both processes and structures, and more theoretical questions about how the OECD exerts influence. In terms of the production process, the authorship of reports and analyses is normally ascribed to the Organisation rather than to individuals, though key names are usually noted in the foreword to OECD publications. This practice reflects the cross-cutting way in which the Organisation operates. Many committees, working parties and expert groups contribute to activities, and many individuals from different sections read and comment on drafts of material. Because of this extensive intra-organisation consultation, all documents are extremely carefully worded and written:

> "There may be fifty to a hundred people in one shape or another who get to see and contribute to the preparation of a document along the way, coming from all sectors . . . It's somewhat cautious and qualified, but if you're tuned in to how to read you can see the punch is there. But the punch would not be hitting people over the head, it would be couched in a way that people could take it and run with it and make an issue of it in the member country, but that it wouldn't cause enormous waves. [This process] probably leads to some blandness in the final text, in the writing styles, because the individual voice gets swamped somewhat . . . the edges get rounded off."
>
> (IT 14: secretariat, 1996)

> "Writing reports is very difficult because there is so much to synthesise and so many different approaches. There's lots of culling and judgment . . . You end up providing a range of ideas and strategies that members can use if they wish."
>
> (IN 33: consultant, 1996)

The low-key style is in keeping with the consensual way the Organisation works, though it is worth noting that OECD reports and analyses are subtly normative. The 'punch' — the Organisation's own stance — is often hidden behind the compilation of comparative country case studies and behind a distanced or disinterested academic style. OECD reports fit the genre of writing Atkinson (1990) classified as "authoritative texts". They are "as much an artefact of convention and contrivance as is any other cultural product" (p. 7). Authoritative texts are 'self effacing' in style, attempting to give the appearance of what Barthes (1967) has called "degree zero" writing. Such a technique seeks to minimise the apparent presence of the author in the text and thus to uncouple the relationship between the author and the subject of the text (Atkinson, 1990, p. 46). Such a style is useful — necessary even — given the divergent interests and viewpoints represented by the Organisation. It is a genre that also represents a high level of generality and abstraction, and is able to smooth over or reconcile such differences. However, 'degree zero' writing in OECD texts also allows the smuggling in of normative stances, which are most obviously implied in the recommended policy options. The OECD's policy stances will be examined more closely in Chapter 3.

OECD's educational work is used, and exerts influence, in widely different ways, depending on the kind of relationship a programme generates (e.g. the institutional basis of IMHE membership compared with the Education Committee's intergovernmental character), the nature of the project, the availability of alternative sources of policy ideas (for example, the European Union), and the size of country. Large 'self-referential' countries such as the UK and the US take relatively little interest in the OECD, it is claimed, while smaller countries are more likely to use OECD ideas as a reference point for education policy decisions.

> "I don't think there's ever been a good relationship between Britain and the OECD."
>
> (IN 26: consultant, 1997)

"The UK is not very interested in the OECD and doesn't take country reviews seriously, compared with the Nordic countries. This is despite the fact that British scholars have made a strong contribution to OECD work."

(IN 29: consultant, 1997)

By contrast, Alan Ruby, former Australian representative on CERI Governing Board and Education Committee (and former Chair of that Committee) argues that Australia participates in the Organisation because:

"It improves our policy making; it is a source of innovation and a stimulus to reform; and it lets us monitor trends and developments in like nations. It does all of these things within an economic and social framework that is compatible with the general direction of public policy in Australia."

(Ruby, 1997, p. 15)

OECD work is disseminated through various vehicles: in books and reports; in unpublished conference proceedings; through academic research papers and reports promulgated at 'meetings of experts'; through reports presented at ministerial and intergovernmental conferences; and more recently, through its web sites. In all of this, it is difficult to track the nature of the OECD's impact or the flows of influence. Perhaps the concept of "a field of interacting forces" (IN 16: secretariat, 1995) is most useful, a point to be developed further in Chapter 3. It is worth noting, however, that despite its prolific output, much of the OECD's educational work appears relatively inaccessible except to relevant government ministries, to a relatively closed network of consultants, or to OECD *aficionados*.

"I think [the OECD] has an influence on the thinking of some of the people who attend the meetings in the sense of if they're senior policy people it gives them a opportunity to test their thinking out . . . I guess I've little sense of it reaching other policy-makers if they're not part of that network . . . I think particular reports or initiatives that become high profile, like the indicators stuff, do reach an awful lot of people at that senior policy-making level just because of the national press calls attention to it."

(IN 28: consultant, 1997)

Further, although its Paris bookshop is a mecca for interested researchers, OECD publications are not routinely available in bookshops and its country-based distributors are not always well stocked. Dissemination is not necessarily seen as a problem though:

" . . . it's not the kind of organisation that's trying constantly to have an influence on people, it's an organisation which provides a means whereby people, key actors and players, can come together and work out their own ideas, collectively . . . So dissemination or impact means something a bit different in that context."

(IT 56: secretariat, 1995)

Nevertheless, driven by the need to assert its place in the new competitive environment of international organisations, the OECD has adopted more aggressive marketing strategies, including forays into cyberspace. In 1997, it established a new Directorate of Public Affairs and Communication, with a role of maintaining:

> " ...   fruitful relations with the press and opinion-makers, while spreading the Organisation's key messages to the general public ...   [and] to key audiences, including parliamentarians, academics, policy analysts, the business community and non-governmental organisations."

> (OECD, 1998a, p. 94)

In its own words, the Organisation's well-recognised concern with "discretion" and "confidentiality" is "no longer [seen as] appropriate in a complex world that demands more transparency and quicker, easier access to information". Communicating "differently" and "more effectively" has been one way, the Organisation suggests, of assisting member country governments "to fully explain to their citizens" the complexities of macroeconomic reforms and structural adjustments (OECD, 1999).

Increasingly, the OECD is reaching out beyond its own members. One of the three deputy secretary-general's posts is now designated within the Organisation's structure as responsible for the wide range of activities relating to non-member countries. Significantly, the OECD is part of a growing network of regional, international and supranational organisations and agencies with an interest in education — UNESCO, the World Bank, APEC, and the European Union to name just a few — in which relationships of cooperation and competition exist simultaneously. For example, UNESCO and the OECD cooperate in collecting statistical data, collaborate with IMHE on some training programmes, and run joint activities on educational programmes — for instance, the use of educational space, performance indicators, vocational and technical education and higher education. There is a view that the external context is urging greater cooperation:

> "There's more cooperation with the OECD over the past five years as the world has become more global. No one can remain isolated now. The OECD is now faced with problems of a kind it wasn't used to — it has entered the 'real world' that UNESCO has always had to deal with."

> (IN 23: international organisation, 1995)

> "There's a similar trajectory for education in the OECD and the World Bank. For example [the World Bank's] *Priorities and Strategies for Education* puts forward a very similar range of themes to the OECD."

> (IT 43: consultant, 1996)

At the same time, a lack of cooperation and greater competition between international organisations in the current context has also elicited comment:

> "There are some people [from OECD and UNESCO] working in the same area who have never met ...   for example, the OECD is developing work on Russia without using the high level expertise which exists in UNESCO."

(IN 19: bureaucrat, 1997)

"There is little cooperation between the two organisations. Indeed, there is competition now in Eastern Europe, for example over who does country reviews. For example, Hungary chose OECD not UNESCO. Why?"

(IN 22: international organisation, 1995)

"There has been a conference frenzy since 1992, with different agencies operating in different arenas or via different processes. For example, the Council of Europe, with its concern for democratising higher education, operates within a legislative framework. The EC is concerned with reconstruction, assisted by its huge budget. The OECD, with no money, is trying to promote its liberal economic ways of thinking, essentially about establishing efficient systems for academic mobility. All these international organisations are competing for territory . . . "

(IN 61: international organisation, 1995)

Ideologically, the OECD occupies a position midway between, say, the World Bank and UNESCO, though the ideological stances, of course, do not remain static:

"It's possible to rank international organisations on a continuum, with the World Bank at the hard extreme and UNESCO on the soft extreme. There's a bit of UNESCO trying to move towards the other end . . . The OECD is towards the World Bank side, but trying to move more towards the centre."

(IN 61: international organisations, 1995)

Operationally, the OECD has neither the legal, political or funding clout of the European Union nor the financial clout of the World Bank. Nor do its conference resolutions — in education at least — mandate its member countries as do the conferences of the UN system. Rather, its authority flows from the perceived quality of its reports and analyses, and, perhaps, from its exclusive membership. In many ways, the OECD draws its strength from the 'new international order' which grew out of the Marshall Plan following the Second World War. This may be changing with the genesis of a 'new world order' associated with the phenomenon most often described as globalisation.

The OECD itself is very much aware of the effects of globalisation upon its functions and ways of operating. In the 1997 Annual Report, for instance, the Secretary-General noted:

"The role of the OECD in the new world, the 'global village', is growing. The collapse of communist regimes, the transformation of Asia, the move to market economies, the freer flow of goods and services and greater diffusion of capital and technology, have all been part of the potent recipe producing this more open, 'smaller', globalised world."

(OECD, 1998a, p. 5)

In the changing post Cold War world, the OECD has moved to embrace non-member countries more firmly. In its own words, it is "moving beyond a focus on its own countries and is setting its analytical sights on those countries — today nearly the whole world — that embrace the market economy" (OECD undated, circa 1997, p. 5). New to the fold are the 'economies in transition' in Eastern Europe and the former Soviet Union as well as the 'dynamic economies' of Asia and Latin America. Currently, the Organisation includes among its priorities the maintenance of globally competitive markets, ensuring that the benefits of globalisation are shared across societies, combating unemployment and working towards social cohesion (OECD, 1998a, p. 99). The OECD is both a globalisation actor as well as being acted upon by globalisation. What this means for the Organisation's educational work is elaborated in Chapter 3. In the next chapter, the question of globalisation will be analysed in more depth. We will look in particular at the effects of political globalisation on policy production processes and on priorities in education.

2

Globalisation and Changing Educational Policy

"OECD has evolved greatly in the globalising world economy. It has been 'globalising' itself, notably through new Members and dialogue activities ... Further, analysing the many facets of the process of globalisation, and their policy implications, has become the central theme in OECD's work, as the challenges and opportunities of globalisation have become a high priority of policy-makers in OECD countries."

(OECD, 1996f, p. 15)

"Better communicating the OECD 'policy message' to have a greater influence on policy formulation is of critical importance. A broad consensus exists on many aspects of the policy requirement for a globalising world economy. But policymakers, politicians or the general public are not always convinced, and many misconceptions persist ... It is necessary to identify clearly the necessary targets for OECD's policy message and maximise its influence on policy."

(OECD, 1996f, p. 16)

In recent years, the idea of globalisation has assumed considerable importance in educational thinking. Educational policy makers and theorists alike have sought to understand both the ways in which global processes affect education and the manner in which education must respond to it. For international organisations in particular, globalisation has become a key concept with which to interpret the enormous economic, political and cultural changes that characterise human society at the beginning of the 21st century. The OECD is no exception. As the quotations at the head of this chapter illustrate, it has had to come to terms with the contemporary dynamics of globalisation. This book seeks to describe the OECD's understanding of globalisation, especially as it relates to considerations of educational policy. The scope of this study is, however, a little broader. We are interested not only in examining the ways in which the OECD has understood and utilised the notion of globalisation. We are also interested in analysing the ways in which the processes of the production and dissemination of educational ideas within and through the OECD reflect some of the self same global processes to which it has sought policy responses. This chapter is to examine more generally recent literature that explores the relationships between globalisation and educational policy, both at international and national levels. This is done to provide a theoretical grounding for the more focussed analysis of the OECD and educational policy in the following chapters.

Within studies on policy, several theorists have recognised that various processes of globalisation affect public policy. In terms of both the processes of policy making and its substantive content, however, globalisation remains an under-theorised domain (Deacon et al., 1997, p. 1), even within education. However, studies are beginning to emerge within educational policy analysis which explore the nature of these connections (e.g. Ball, 1998; Henry et al., 1999). They have addressed linkages between two broad areas, namely, the linkages between globalisation and educational restructuring and those between globalisation and a new educational policy consensus (Brown et al., 1997, p. 7). The former fits within the theorising of new state structures and a new managerialism, which has affected educational policy production in ways similar to its impact upon the production and practices of new health and welfare policies. The latter looks at the new human capital framing of educational policy (e.g. Marginson, 1997b) and the attempts to create supposedly post-bureaucratic state educational systems which in varying degrees embrace quasi-market approaches (Whitty et al., 1998).

A number of scholars have in recent years asserted the need to develop new theoretical perspectives around the linkages between globalisation and new production processes and content in educational policy. Deacon et al. (1997, p. 2), for example, argued that the increasingly nomadic, highly mobile, global capital has reduced the policy salience of governments at the nation state level. A disjunction has emerged between the spatial frame of the nation and that of nomadic capital within the global economy, particularly in respect of financial markets, facilitated by technologies which effectively annihilate time and distance. In contradistinction to those who argue that an emergent post-national and supranational politics has led to the demise of the nation-state, our position is that the nation-state remains very important for policy, but that it has been reconstituted in a number of important ways by globalisation. It is the flowback between globalisation and the reconstitution of the nation-state and its policy producing structures in education which provide the focus of this chapter.

This chapter focusses in upon political globalisation, given the salience of understanding the changing global political arena in analysing the contemporary workings and policy emphases of the OECD itself. An account of political globalisation is also necessary to get a better grasp over both the reconstituted policy processes in the nation state and over changing relationships between the OECD and member nations and their educational policy regimes. These processes are located within the various political, economic and cultural flows and effects of globalisation. Some of these issues are also pursued in Chapter 3.

This chapter looks first at the reconstitution of the nation-state set against these flows. The impact of the globalisation of the economy and dominance of neo-liberal ideology are analysed next, as is the latter's manifestation in a post-Keynesian public policy consensus and its consequences for educational policy within nation-states. Together with the global dominance of neo-liberal ideology in the post Cold War world, geo-politics has developed a view about the organisational form the state requires if it is to ensure global competitiveness for the national economy (Yeatman, 1998). In this setting, Waters (1995, p. 80) coined the notion of 'organisational ecumenism' to explain the global dominance of a particular organisational form across both public and private sectors. These new state structures framed by 'corporate managerialism' (Considine, 1988) or 'new public management' (Hood, 1995) and by new forms of governance will then be considered, as well as their significance for educational policy. Finally, the chapter will summarise the argument and will put forward the view of the emergence of a global educational policy community of which the OECD is a part. It will simultaneously emphasise the continuing significance of the state at national and sub-national political levels for educational policy production. Throughout, attention will be paid to the significance of the various cultural flows within globalisation. They are facilitated by new technologies and by those forms of instantaneous communication which compress time and space (Harvey, 1989; Giddens, 1994). Likewise we will attend to the cross-national flow of policy concepts and to the emergence of a global policy community.

Reconstituting the Nation-State

Globalisation creates pressures upon the nation-state from above and from below. Commenting on the moves to devolution within the UK — with its limited forms of parliamentary independence for Scotland and Wales — Giddens (1998) observed that, "Globalisation not only pulls upwards, it pushes downwards, creating new pressures for local autonomy". Such destabilisation and reconstitution of the nation-state from both above and below affects policy-making processes and the available policy options for governments.

Structurally, a number of interrelated developments from above, the consequence of globalisation, affect the policy salience of the nation-state. These include: the enhanced globalisation of the economy; the related extra-territorial character of global capital; the post Cold War context; and the apparent global dominance of neo-liberal ideologies. In combination, such developments have weakened the policy options of nation-states. But they have also created some post-national political structures, as well as reconstituting international organisations already in place. The emergence of the supranational political unit of the European Union, new regional trade zones such as the Asia–Pacific Economic Cooperation forum (APEC) and the North American Free Trade Agreement (NAFTA), are good examples of this dynamic. The European Union is the most advanced of the supranational political units with member nations delegating some of their legislative authority to this unit beyond the nation. Politically, this process served to soften the new right political agendas pursued by Thatcherism in the UK and to weaken the social democratic agendas of certain continental European member nations. It also laid down clearly defined criteria for economic policy for member states in relation to the single monetary union. New regional entities such as NAFTA and APEC, while at present functioning largely as economic and trading arrangements, also place policy limitations in these matters upon their members. Such limitations are most obvious when related to trade liberalisation and to the removal of tariff protection to ease the move towards more permeable national boundaries and a borderless global economy. The rise of such an economy in turn limits the policy salience of the nation-state and its room to manoeuvre independently.

One of the factors in the destabilisation of the nation-state is the growth of so-called 'region states' (Ohmae, 1995), examples of which would be Vancouver and Seattle or Toronto and Cleveland, athwart the Canadian–American frontier. Region states cut across the sovereignty of nation states. They work almost as separate economic units. As "economically integrated regional entities, ... [their] primary linkages tend to be with the global economy and not with their 'host' nations" (Axtmann, 1998b, p. 4). Compounding this have been the end of the Cold War, the rapid mobility of capital, and the emergence of global cities or mega-cities (Castells, 1996). Almost cut off from their host nations, they provide a number of world economic centres — a polycentric situation different from both that of the Cold War period and from an earlier colonial era. Mega-cities are spreading within developing countries, for example Shanghai and Beijing in China, while 'Third World' populations are emerging in First World nations. Such trends illustrate the growth in inequality within nations as well as between nations, which is associated with the predominance of neo-liberal economics.

The recognition of the global nature of contemporary political problems and political processes has brought forth a proliferation since the 1960s of international Non-governmental Organisations (NGOs) such as Greenpeace or Amnesty International. These sit alongside the substantial number of already existing Intergovernmental Organisations (IGOs), such as the UN system, the World Trade Organisation and the World Bank, whose relationships were systematised in the aftermath of the Second World War. Waters (1995, p. 113) has argued that, "Together they [IGOs and NGOs] constitute a complex and ungovernable web of relationships that extends beyond the nation-state". The thesis of global 'ungovernability' is an important aspect of the reconstitution of the nation-state and its remit in public policy production.

In addition to such relationships existing above and beyond the nation state, there has been fragmentation within, as well as the break-up of, some larger political units, for example, Yugoslavia, the Soviet Union, and Czechoslovakia. Small states have also emerged — states, which often seek to reassert the unity of ethnicity and territoriality, a unity which has also been torn asunder by the flows of globalisation. Bauman (1998) argues that these small states are a concomitant of globalisation:

> "The rush to carve out new and even weaker and less resourceful 'politically independent' territorial entities does not go against the grain of the globalising economic tendencies; political fragmentation is not a 'spoke in the wheel' of the emergent 'world society', bonded by the free circulation of information. On the contrary — there seems to be an intimate kinship, mutual conditioning and reciprocal reinforcement between the 'globalisation' of all aspects of the economy and the renewed emphasis on the 'territorial principle'."

(Bauman, 1998, p. 67)

Appadurai (1996) takes the view that we have entered a stage of postnational politics, or at least we are seeing the inchoate beginnings of the dissociation of politics from the territorial space of the nation. Migration and media taken together have sewn the seeds of postnational politics, so this line of argument suggests. In this setting the concept of media serves as a shorthand for various advances in technologies (and forms of transportation) which compress time and space and act as constraints upon communications. Agreed, there is a greater interconnectedness across the globe of economics, culture and politics. Actions at a distance affect the local (Giddens, 1994). The micro-histories, economies and cultures of the local are catalysed by global pressures to produce fragmentation and difference within nations. Thus the impacts of the global upon the local are differentiated within nations, as is the case with the clear spatial location of poverty and the related rise of backlash politics of various kinds.

As Axtmann (1998a, p. 2) has noted: "Arguably, a reference to 'globalisation' contains the hypothesis that there has occurred an increase in the density of contacts between locations worldwide", an observation that agrees well with Waters' definition of globalisation as a "social process in which the constraints of geography on social and cultural arrangements recede and in which people become increasingly aware that they are receding" (1995, p. 3). Marginson (1999, p. 24) suggests that it is in the phenomenal world of individuals that globalisation has gone furthest. In many other respects, even including the economic, it is a "radically incomplete" project. It is in such conditions that a rift develops within nations between the globalised and globalising elites, and those who have been adversely affected by globalisation through unemployment, job insecurity and growing inequality. In Australia as elsewhere, globalisation is witness to the growth of two nations, between the political elites of the global cities and two other groups, namely country dwellers disadvantaged by low levels of education and the decline of commodity prices, and the outer suburban working (and non-working) classes, affected by the sharp decline in the employment capacity of manufacturing industries. Martin and Schumann (1997) refer to the emerging 20/80 society clearly divided between a globally connected elite (20%) and a subordinated layer (80%), whose connections with the global economy are only through consumption behaviour and the media.

Bauman (1998) holds the cleft between the globally mobile and the immobile constitutes a new dimension of inequality within the rich nations of the world. As he felicitously puts it: "Mobility and its absence designate the new, late-modern or postmodern polarisation of social conditions. The top of the new hierarchy is exterritorial; its lower ranges are marked by varying degrees of space constraints, while the bottom ones are, for all practical purposes, *glebae adscripti*" (p. 105). This polarisation within societies poses a considerable challenge to the policy solutions available to the nation-state and speaks to a divide between elites and others. Today, the metropolitan elites of the mega-cities of the world often have more in common with similar elites in other such cities than with the immobile mass within their own nations — the 'victims' of globalisation, or of the global dominance of neo-liberal economic ideology.

Against this backdrop Appadurai (1996) developed the proposition of there being an inchoate postnational political challenge to the 'nation-state', where the hyphen between nation and state has faded somewhat. This concept of scapes — ethnoscapes, mediascapes, technoscapes, financescapes, ideoscapes — depicts global cultural flows and explores the disjunctions between economy, culture and politics that result from globalisation (p. 33). Postnational politics emerge through the scapes and flows of globalisation. Cultural scapes in Appadurai's terminology embraces the flows of people (e.g. migrants, refugees, tourists, students, politicians, intellectuals, policy elites) and the rapid flow of images and ideas via new technologies. Peoplescapes, including the flows and communications between politicians, policy elites, policy intellectuals, and other players within the educational policy community, reduce the distance between elites and intensify the frequency of their communications. In this vision, we see the eventual emergence of a postnational policy community consisting of top-level bureaucrats and politicians, policy makers, policy advisers, and policy intellectuals. This is both a form of influence and a new context of influence (Bowe et al., 1992) in the policy cycle. It sits above the nation-state and its policy producing apparatus. In Appadurai's construct, ideascapes refer to the rapid flow of ideas, in this case, of educational policy options which contribute to a diaspora of key policy options and concepts. Technoscapes facilitate both communications between policy elites and the flow of policy ideas.

Appadurai argued that ethnoscapes — the rapid flow of peoples across the globe — combined with technoscapes leads to the creation of what he termed "diasporic public spheres". Diasporic public spheres exist when migrant groups living in the new host nation continue to participate in the politics of their country of origin through the use of the internet, telecommunications, media programs, and rapid transportation. Appadurai holds these diasporic public spheres to be embryonic forms of postnational politics; "Diasporic public spheres, diverse amongst themselves, are the crucibles of a postnational political order" (Appadurai, 1996, p. 22). The important flows associated with globalisation which are incubators of a postnational political order, he asserted, mean that today "new ways [exist] in which individual attachments, interests, and aspirations increasingly crosscut those of the nation-state" (p. 10). While Anderson (1983) examined the "imagined community" of the nation, Appadurai extended this concept as it applies to postnational politics and diasporic public spheres seeing it in terms of "imagined worlds" (p. 33) which bring about the disjuncture between nation and state.

While we are more sceptical than Appadurai about the demise of the nation state, we certainly recognise the emergence of postnational politics. Appadurai also suggested it might be that "the emergent postnational order proves not to be a system of homogeneous units (as with the current system of nation-states) but a system based on relations and networks between heterogenous units (some social movements, some interest groups, some professional bodies, some non-governmental organisations, some armed constabularies, some judicial bodies)" (Appadurai, 1996, p. 23). The concurrent workings of those units with the politics of the nation state contribute to the latter's reconstitution and on a global scale give rise to the era of apparent ungovernability.

The changed geo-politics of the post Cold War world has also contributed its part to this situation. Bauman (1998, p. 58) suggested that the end of the "Great Schism", as he calls it, created a world which "does not look like a totality anymore". Rather, he remarked, the world now looks "like a field of scattered and disparate forces, congealing in places difficult to predict and gathering momentum which no one really knows how to arrest" (p. 58). No one appears, Bauman suggested, to be in control. Indeed, it would be difficult to see what the very concept of 'being in control' might mean in these new globalised circumstances. These forces can be juxtaposed with the central pervading belief of modernity that the nation-state could control and manage the affairs, economic, political and cultural, within its boundaries. Bauman notes that we tend to speak of the effects of globalisation rather than conceiving of some central agencies driving globalisation in a coherent manner. Indeed, regarding the very definition of globalisation, he observes: "The deepest meaning conveyed by the idea of globalisation is that of the indeterminate, unruly and self-propelled character of world affairs; the absence of a centre, of a controlling desk, of a board of directors, of a managerial office" (Bauman, 1998, p. 59).

In contrast, the modernist construction of the nation-state rested upon an assumption of national sovereignty relating to the military, economy, culture and politics. As this, and subsequent, sections seek to demonstrate, the nation-state no longer has autonomous sovereignty over these matters, particularly the economy. The capacity for Keynesian style demand stimulation has been reduced, at least given the globally dominant neo-liberal ideology. Bauman (1998, p. 65) interprets globalisation's challenge to the policy capacities of the nation state in the following way.

> "Any control of such 'dynamic equilibrium' is now beyond the means, and indeed beyond the ambitions, of the overwhelming majority of the otherwise sovereign (in the strictly order-policing sense) states. The very distinction between the internal and the global market, or more generally between the 'inside' and the 'outside' of the state, is exceedingly difficult to maintain in any but the most narrow, 'territory and population policing' sense."

> (Bauman, 1998, p. 65)

Wiseman (1998, p. 14) suggested that "the processes of globalisation are helping to create a world of 'nested locales' in which households, neighbourhoods, cities, provinces, nations and regions sit inside the wider global relationships like Russian Babushka dolls." The destabilising effects upon the nation-state flow from the fact that relationships between these nested locales are two-way, both top-down and bottom-up, and consequently reconstitute the political domain and thus the policy capacities of the nation-state.

These relationships highlight the more polycentric nature of politics today. Extending this further, Holton (1998, pp. 132–133) describes global politics as a multicentred 'cobweb' of relationships as opposed to simply politics between nations. This 'cobweb' embraces two-way interactive relationships between government and international governmental and non-governmental organisations, as well as between bodies such as these economic units (both local and global) and political and pressure groups of civil society both within and outside the nation. The result is an extended policy community in any policy domain, which includes issues networks (Rhodes, 1997), together with dual local and global pressures which work to reconstitute the nation-state.

The distinctions between the empirical and ideological aspects of globalisation — that the ideological aspects also have empirical effects — deserves further attention. For example, post-Keynesian, neo-liberal policy responses have reduced the policy ambitions of nation-states, yet one should avoid falling into the "myth of the powerless state" (Weiss, 1997) and accepting that the sole role for the nation state today is to facilitate economic globalisation. Even though globalisation has destabilised the nation-state, with politics now operating within and across nations, supranationally and intranationally, the sustained significance of the reconstituted apparatus of state in politics and policy production should be taken into account (cf. Dale, 1997). Perhaps the greatest destabilising influence on the sovereignty of the nation state is the global economy, an issue which is discussed in the next section.

Globalisation of the Economy and a Post-Keynesian Policy Consensus

The globalisation of the economy has had considerable impact upon nation-states. Hobsbawm (1994), in characterising the global economy, drew an important distinction between international and transnational dimensions of economic activity. The latter refers to "a system of economic activities for which state territories and state frontiers are not the basic framework, but merely complicating factors" (Hobsbawm, 1994, p. 277). Regarding the creation of a global economy, he adds:

> "In the extreme case, a 'world economy' comes into existence which actually has no specifiable territorial base or limits, and which determines or rather sets limits to, what even the economies of very large and powerful states can do. Some time in the early 1970s such a transnational economy became an effective global force."
>
> (Hobsbawm, 1994, p. 277)

Such a world economy destabilises the nation-state, which in its modernist form had responsibility for the well being of its national economy. Now, nomadic, footloose global capital moves beyond the purview of the nation-state and national policies. This is a tendency rather than a foregone conclusion because it has also to be recognised that such capital requires social and political stability (cf. Holton, 1998), which can only be assured by policies operative within nations. The global economy exists in its most mature and postnational form in financial markets, with trade also moving in this direction, sponsored by the pervasive trade liberalisation policy hegemony of most of the major international organisations such as the World Bank, the International Monetary Fund, the World Trade Organisation and the OECD. Integrated global financial markets which move billions of dollars daily through electronic transfers have substantial effects which are reflected in national economic indicators and economic policies of national governments, including "foreign exchanges, interest rates, the stock market, employment levels, and government tax revenues" (Holton, 1998, p. 80).

Against this background, governments in the rich countries of the global economy have pursued neo-liberal economic polices and rejected state interventionism and demand management developed during the period of Keynesianism which stretched from 1945 until the mid-seventies. Certainly, this is an ideological reading of globalisation, which in turn has real consequences. Given the global mobility of capital as opposed to the territorial 'groundedness' of labour and national governments, the policy options for governments are somewhat limited. This limitation stems from the fact that, unlike the Keynesian period, the nation-state is today less able to maintain itself in economic isolation from the global economy. This constraint works in different ways for economies of varying sizes. Many transnational corporations today have economies larger than those of mid-range nation-states. Of the world's one hundred largest economies, forty-seven are multinational conglomerates (Latham, 1998, p. 11). Furthermore, the volume of daily currency trading dwarfs the economic capacities of most nation-states and creates inhibitions for most nations regarding policy options.

With the globalisation of the economy and support for neo-liberal ideology by international organisations and by nation-state based political parties, most nations have sought to 'globalise' their economies or, alternatively, attempted to ensure the global competitiveness of their national economies. Such a goal has taken on meta-policy status (Yeatman, 1990) and as such frames the nature of other policy domains, including education. Pusey (1991, pp. 210–211) argued that these policy goals and framework "presuppose a closer functional incorporation of the 'political administrative' system (the state, and with it the obligatory conditions of elected governments) into an augmented economic system". Similar views are found in Habermas (1996, p. 292) who observed: "While the world economy operates largely uncoupled from any political frame, national governments are restricted to fostering the modernisation of their national economies." Nation-states have basically become "partners to global economic players" (Marginson, 1999, p. 26).

The political concomitant of globalisation of the economy has been the dominance of neo-liberalism, replacing the post World War II era of Keynesianism and taking the form of a post-Keynesian policy consensus. This consensus gives primacy to the market over the state as a societal steering mechanism, supports balanced budgets and works to reduce government expenditure and government intervention. Market approaches to policy delivery have also been taken up by the state, whose structures have as a result been substantially restructured. This aspect will be addressed in detail later in this chapter.

'Leaner and meaner' the state now operates against a background of growing social instability, loss of social cohesion and deepening inequality. Yet, the new policy consensus fails to offer measures to counter the perverse effects of the neo-liberal espousal of globalisation. The enhanced flows of people — ethnoscapes — which accompany globalisation, ensure that the populations of nations are more ethnically diverse. Multiculturalism, which Fraser (1995) regards as a form of "recognition politics", becomes an issue within nations. Policies undertaken by governments vis à vis recognition politics — anti-discrimination, affirmative action, anti-vilification laws and so on — in an atmosphere of growing inequalities and pervasive insecurity, have spawned a backlash, and generated political resentments and new racisms evident in the so-called culture wars within nations (McCarthy, 1998). The rise of xenophobic movements such as One Nation in Australia, Lepenism in France, white working class militia movements in the US, and neo-Nazi groups in Germany and the UK, show that the management of ethnically diverse populations has become an important issue for governments, in which education plays a particularly important role.

Bauman (1998) sketches a depressing account of the effects on public policy of the quasi-symbiotic relationship between economic globalisation and neo-liberal ideology. The falling demand for labour in a vastly reduced state is parallelled by a growth in levels of redundancy, substantial increases in imprisonment rates and the virtual criminalisation of poverty. To the pervasive insecurity accompanying our "manufactured uncertainties" (Giddens, 1994) governments respond with policies of law and order rather than facing up to the causes behind the situation. Indeed, the new policy consensus prevents steps being taken that might ameliorate the conditions of inequality and resentment. Bauman (1998) argues that disciplining populations is one option still available to national and sub-national governments and, moreover, is demanded by the insecure and anxious faced by the depredation of economic globalisation. He summarises his analysis in this strikingly, hyperbolic fashion:

> "To focus locally on the 'safe environment' and everything it may genuinely or putatively entail, is exactly what 'market forces', by now global and so exterritorial, want the nation-state governments to do (effectively barring them from doing anything else). In the world of global finances, state governments are allotted the role of little else than oversized police precincts; the quantity and quality of the policemen on the beat, sweeping the streets clean of beggars, pesterers and pilferers, and the tightness of the jail walls loom large among the factors of 'investor confidence', and so among the items calculated when the decisions to invest or de-invest are made."
>
> (Bauman, 1998, p. 120)

There is clearly much that is weighty in Bauman's account. Yet, it sits in a framework that suggests the nation-state, faced with globalisation, has little political and policy room for manoeuvre. Basically, the nation-state has to ensure the international competitiveness of its national economy. Such an imperative limits its other options and feeds a "pathology of diminished expectations about democracy" (Hirst and Thompson, 1996). In the current conjuncture, a variety of political options exist, however, along a continuum between social democratic and New Right manifestations of post-Keynesianism. Examining the former, Giddens (1998) has argued for the need for a social democratic Third Way, to reconstitute, revitalise, and rearticulate social democratic politics in the face of globalisation. The Blair Labour government in the UK has attempted to pursue this so-called Third Way between old style state centred democratic socialism identified with Keynesianism and the New Right market society associated with Thatcherism. Preoccupation with social justice has been revived through the issue of social exclusion (Lister, 1998), as it has been throughout the European Union (Levitas, 1996; Brine, 1999). National Labor governments in Australia (1983–1996) also attempted to tie in new market approaches with redefined notions of social justice (Lingard et al., 1993). In the aftermath of the electoral defeat in 1996, some Labor politicians in Australia (Latham, 1998; Tanner, 1999) have urged the need to rethink social democracy, stressing the importance of governments for the creation and support of both social and economic capital to uphold social cohesion and to reduce inequalities. Here too education policy is held to be central to the formation of social capital and to the democratic spread of social capabilities.

Some of this rethinking recognises that economic growth and development are impeded by social dislocation, by growing inequalities and by the related collapse of confidence, social and civic. Garrett and Mitchell (1996) argued that more productive and redistributive policies and approaches to welfare do not necessarily deter potentially mobile capital because social stability also attracts such capital. Some global capitalists — George Sorros being one — are now arguing that social cohesion, stability, not to mention both global and nation-state based regulation, are necessary to the future of global capitalism. It is an argument broadcast by some international organisations in their predictions about the future of the global economy. Here perhaps a new window is beginning to open for a rethink of social democratic politics in the face of globalisation, a point to be developed in the concluding chapter.

Educational policy has been reframed by the new policy consensus resulting from the combination of globalisation and neo-liberal ideology. Within the sociology of education it is generally agreed that an earlier policy consensus accompanied the post-war period of Keynesianism (Brown et al., 1997), which endured until the mid-seventies. During that period, education was central to the pursuit of two policy goals, namely, growing economic prosperity and equality of educational opportunity for all. The former was framed by a macro-economically focussed human capital theory, while both were to be pursued through increased educational expenditure. The organisational form of educational systems followed the model of classical bureaucracy as Weber defined it. The policy approach during the period of economic nationalism was the outcome of a number of political pressures, from governments for human capital building for economic ends and from civil society for an expansion of opportunities and a more equal and cohesive social covenant (Marginson, 1997b, 1999). As has been argued earlier, globalisation broke down economic nationalism which in turn ushered in the post-Keynesianism era as it has been described up to now. The new educational policy consensus fits within this framing. Essentially, it involves the coming together of the effects of globalisation and the dominance of neo-liberal ideology that gives rise to the meta-policy within nations of ensuring an internationally competitive economy and with it a new consensus in educational policy. Further, and as this book demonstrates, international organisations such as the OECD have served as significant mechanisms for institutionalising this new consensus (Spring, 1998).

Within the global economic framework, education is now regarded as the policy key to the future prosperity of nations (Brown et al., 1997, pp. 7–8). This programme has been presented as a new human capital theory, focussing on micro-economics, which stress the importance of a highly skilled and flexible workforce to national success within the new global knowledge economy (see Drucker, 1993). Marginson (1999, p. 29) has pointed out that: "The augmentation of skills and knowledge (including management capacity) are means of coupling units of 'footloose capital' to particular populations in particular nations."

In education, the new human capital approach is regarded as much an individual benefit as a social one. Earlier educational policy wisdom viewed education as a social good which justified increased funding. Redefining education as an individual good justified introducing the principle of 'user pays' in education, a principle best seen in the fee contribution scheme for university students in Australia. Neo-liberal ideology has introduced quasi-market approaches and new post-bureaucratic state educational systems, dimensions which will be developed in the next section. Though education is now deemed more important than ever for the competitive advantage of nations, the commitment and capacity of governments to fund it have weakened considerably. As Marginson (1999, p. 30) observed: "Thus educational institutions are more politically central than ever, yet they also seem to be weakened, fallen from a former high estate, hostage to the diminishing of the nation-state's room to move and its growing fiscal constraints." However, new issues and the creation of social capital are beginning to find their way onto the policy agenda. Issues such as the relationship between social stability and economic productivity are emerging, as social democracy tries to adjust to globalisation.

New State Structures and Forms of Governance

The transition from multinational to transnational capitalism during the sixties and seventies resulted in uncoupling the 'needs' of such a form of capital from the organisational structures of the state. It did so at both national and sub-national levels (Yeatman, 1998). From this transition new state structures and new forms of governance evolved. Reform in the public sector saw structures and practices of departments and agencies transformed under the rubric of 'corporate managerialism' or 'new public management'. This transformation involved the assimilation of private sector management practices by the public sector. Weight has shifted from an earlier bureaucratic stress on due process — a fetish for proceduralism or 'red-tapeism' (Yeatman, 1990) — to outcomes achieved at the lowest possible costs. The twin goals of greater efficiency — doing things at the lowest cost — and greater effectiveness — achieving the goals set — permeated the new structures which are less hierarchical and flatter, with greater management prerogative for policy steering. Relations between the strategic centre of the organisation concerned with policy and the practice-based periphery have been rejigged, a process often described as 'steering at a distance' in analysing this new centre/periphery relationship (Kickert, 1991). The centre establishes the strategic plan and desired policy outcomes for the organisation. The policy-practising periphery is responsible for achieving these goals. However, any new autonomy at the periphery operates in relation to means rather than policy ends, for these are set tightly by the centre as part of a new regime of outcomes accountability.

Over and above new relations between centre and periphery, relationships between ministers and senior managers have likewise changed, the latter now most commonly employed on performance related contracts, an arrangement which has facilitated the rise of 'ministerialisation' (Knight and Lingard, 1997) or 'politicisation' (Ball, 1990) of policy making. Davies and Guppy (1997) have developed a similar line of thought. The effects of globalisation upon the administrative structures of nation-states, have made educational policy production "too important for educators" and as a result educational policy framing occurs at a "higher level". Simultaneous with this is a 'squeezing' of power from the middle (p. 438). The upshot, they suggest, is "a simultaneous centralisation and devolution of authority that squeezes power from middle levels of educational administration and redistributes it upward to more central states and downward to individual schools and reform groups" (p. 439). Politicians have sought to reassert control over the bureaucracy and the setting of policy agendas. New managerialism was designed to facilitate this drive. The result: policy objectives become narrower, are set at a higher level, whilst responsibility for their achievement is handed off to policy practitioners at the point where the service is delivered. Managerialism has also concentrated on achieving more across-government policy coherence, an important factor in the human capital interpretation of education as a significant element in national economic strategies. Broad policy settings have been framed at a higher level and in the process often exclude professional educational advice.

This administrative re-engineering emerges in different ways in different nations, depending on history, political structures and cultures, and political party in power. In Australia, for example, recourse to statutory authorities operating at some distance from ministers and governments in policy production has been severely limited. In the UK, in contrast, statutory authorities or agencies, e.g. the Teacher Training Agency and Office of Standards in Education, have proliferated in ways which appear to remove responsibility for the outcomes of policy from the relevant ministers. Ball (1998, p. 125) suggests that such "non-interventionary interventions" appeal to politicians because they appear to separate "the reformer from the outcomes of reform", while still allowing policy to be steered.

National (and sub-national) variations in restructuring share common motivations and ideologies. The OECD has been an important vehicle for setting the new managerialism and new forms of governance in place. The OECD report, *Governance in Transition: Public Management Reforms in OECD Countries* (OECD, 1995c), exemplifies the OECD genre of disinterested academic description attached to mild exhortation to change. It demonstrates the pervasiveness of new public management in OECD countries. It notes in a not too muted criticism of state bureaucracies that 'highly centralised, rule-bound, and inflexible organisations that emphasise process rather than results impede good performance' and that the efficiency of the public sector 'has a significant impact on total economic efficiency' (OECD, 1995c, p. 7). Summarising the nature of public sector changes across the OECD, the report says:

> "The common agenda that has developed encompasses efforts to make governments at all levels more efficient and cost-effective, to increase the quality of public services, to enable the public sector to respond flexibly and more strategically to external changes, and to support and foster national economic performance."
>
> (OECD, 1995c, p. 7)

In educational policy, two sharp contradistinctions between bureaucratic educational systems and the new managerialism are set forth to legitimate current changes. Ball pictures the differences thus:

> " ... the grey, slow bureaucracy and politically correct, committee corridor-grimness

of the city-hall, welfare state as against the fast, adventurous, carefree, gung-ho, open-plan, computerised, individualism of choice, autonomous 'enterprises' and sudden opportunity."

(Ball, 1998, p. 124)

Education systems have been substantially affected by new managerialism. Indeed, by implementing it right across the entire public sector, education systems have lost their *sui generis* character. Organisation, structures and basic practices look similar in education, health, welfare and other public sector bureaucracies. It is an outcome achieved through, and expressed in some cases by, appointing career managers to the various public sector bureaucracies. In a sense this new arrangement is post-bureaucratic and follows on the attempt to bring state structures and practices into 'sync' with the needs of global capital (Yeatman, 1998).

Public sector restructuring in education centralises the setting of policy and devolves the responsibility to achieve goals set at the centre, a process which has occurred at both national and sub-national political levels. The capacity to set strategy and accountability at the centre and the devolution of the means to achieve goals and to meet accountability demands constitutes a mix of centralising and decentralising pressures. However, as with the new state structures generally, the specific patterns this restructuring imposes on different systems in different nations are grounded in local histories and cultures. In Australia, for example, at the peak of the progressive Keynesian settlement of the mid-seventies, moves to devolution were also visible, but as a social democratic programme as opposed to the current managerialist vision (Rizvi, 1994). It went hand in hand with a massive increase in educational expenditure and an extensive range of programmes of positive discrimination aimed at advancing equality of educational opportunity and achieving more equal outcomes. By contrast, the newer versions of devolution have been shaped by corporate managerialism, and by market ideologies, the latter being most apparent during the Thatcherite reforms in England and Wales which set schools in competition with each other through the publication of league tables of performance. Attempts were made to break 'provider capture', that is the belief that schools were run by teachers in their own interests. Whitty et al. (1998, p. 3) nicely encapsulated the broad character of the recent wave of world-wide school-based management reforms in the following way:

> "The past decade has seen an increasing number of attempts in various parts of the world to restructure and deregulate state schooling. Central to these initiatives are moves to dismantle centralised educational bureaucracies and to create in their place devolved systems of education entailing significant degrees of institutional autonomy and a variety of school-based management and administration. In many cases, these changes have been linked to enhanced parental choice or an increased emphasis on community involvement in schools. School policy initiatives often introduce a 'market' element into the provision of educational services even though they continue to be paid for largely out of taxation."

(Whitty et al., 1998, p. 3)

Structural arrangements similar to those in schooling have been introduced in higher education. 'Steering at a distance' relationships were laid down between governments and individual institutions. In many OECD countries universities have to respond to portfolios of performance indicators, and provide research profiles as one basis for determining levels of public funding. Institutions have been pushed to raise funds from sources other than government through the sale of research and consultancy services and through the introduction of fees, particularly for full fee-paying international students, who constitute another of the flows of people associated with globalisation. In Australia, the federal government introduced fees for home students on a limited basis as a part of the policy of raising the numbers of university students at manageable cost to the government. These entrepreneurial and fund raising developments in universities challenge their more traditional goals of scholarship, social critique, disinterested pursuit of knowledge and curiosity-driven research. Student fee structures have also modified student /staff and student/institution relationships towards a more consumerist culture.

Associated with new forms of accountability in educational systems is a new culture of 'performativity' (Lyotard, 1984), which ties in with proof of policy outcomes as an important element in the engine of 'remote steering' within the restructured state. The culture of 'performativity' pervades restructured educational systems through imposing a mass of performance indicators linking strategic plans of the centre (the policy producing arm of the restructured state) with outcomes of practice at the periphery (individual institutions). Performativity also relates to deeper epistemological changes which in turn relate to critiques of Euro- and phallo-centric worldviews. They accompany what some see as postmodernity. As a consequence, there has been greater recognition of the perspectival character of all knowledge claims which flow from such post-colonial and feminist theorising (Bhabha, 1994; Harding, 1994, 1998; Kumar, 1997). Yeatman (1994, p. 106) notes that we now have 'no operative consensus concerning the ultimate or transcendental grounds of truth and justice'. One result for state policy practices of the foregrounding of a culture of performativity has been the weight placed on the instrumental, operational and measurable rather than on claims made on the basis of truth. "Performativity is a systems-orientation: instead of the state appearing as the enlightened and paternal command of shared community, the state is equated with the requirements of a system for ongoing integrity and visibility" (Yeatman, 1994, p. 110). Performativity shifts the emphasis to the operational and measurable, thus facilitating the choice of the state between multiple and competing claims on policy in a period of financial squeeze. Yeatman argues that performativity functions as a "principle of selective closure in respect of the information overload and social complexity" (p. 117) with which the contemporary state has to deal.

Within the new forms of governance at national and sub-national political levels in education performativity augments the rationalisation and technicisation of policy. Davies and Guppy (1997, p. 436) develop the concept of "global rationalisation" as another element contributing to the apparent convergence of educational policy structures across nations. Our interpretation holds that this has been augmented considerably by the performative culture now endemic to state policy producing structures in nations across the planet. While the new structures are in some ways post-bureaucratic, performativity increases the apparent rationalisation and flows of information in ways that appear to be more bureaucratic still. Davies and Guppy draw on the work of John Meyer and his colleagues at Stanford University (Thomas et al., 1987; Meyer et al., 1992a, 1992b, 1997) to suggest that the emergence of a world cultural system through the imperialist advance of an organisational form and related practices, operates independently of economic and political globalisation and in fact preceded them. We believe that political globalisation and the new human capital policy consensus in education push such global rationalisation even further through performativity. This line of argument is pursued in Chapter 5 in relation to the development of educational indicators.

The new state structures and cultures are often combined with the quasi-marketisation of educational services (Marginson, 1997b). They are sometimes encapsulated in the idea that a new form of governance is shaping up in OECD countries (Rhodes, 1997). Governance is a concept broader than government. It refers to the various ways in which the historic state/civil society, public/private divisions have been eroded by these structural changes (Popkewitz, 1996; Dale, 1997), and the way structures and practices of government have been reordered. Rhodes (1997, pp. 46–47) suggests that competing usages of the notion of governance refer to the changes in government which occurred following restructuring to meet globalisation. These competing usages include: the minimal or hollowed out state, corporate governance, new public management, good governance, government as a socio-cybernetic system and as self-governing networks. According to Rhodes governments have three basic channels for steering policy, namely, traditional hierarchical bureaucracy, markets or networks. Given the merging of public/private distinctions which flowed from, and are inherent in, market versions of the new managerialism, Rhodes argues that governance depicts "the self-organising interorganisational networks" which characterise government today (1997, p. 57). Examples of these new modes of governance are the involvement of private sector organisations in the provision of welfare services, or the joint private and public sector involvement in the Blair government's development of Educational Action Zones, aimed at overcoming educational inequality in some of the most disadvantaged localities in England. Governments, according to Rhodes, set the policy parameters for particular policy networks, networks that may extend through new technologies to include global actors and ideas. This new form of governance — of "government without governing" as Rhodes puts it — may be seen as another aspect of the destabilisation of the nation state and its policy capacities from within and below.

Governance may encompass the concept of governmentality of Foucault (1991), which refers to the technologies and discourses of government which seek to produce self-governing individuals and subjectivities. Rose (1990) writes of such technologies as desiring to "govern the soul". The neo-liberal state reforms analysed in this section have resulted in new technologies and discourses of government, particularly a 'performative' culture. They are concerned with producing particular subjectivities of state workers and of those in receipt of state policies. Blackmore (1999), for example, examined how performativity within higher education reconstituted individual behaviour. Individuals spend as much time being seen to perform as actually performing, and hence they govern themselves. Steering at a distance and supposed greater autonomy within strategic frameworks for practitioners does not suggest that the state has foregone its competence. On the contrary, as Rose and Miller (1992, p. 174) note, such practices are "not the antithesis of political power, but a key term in its exercise, the more so because most individuals are not merely the subjects of power but play a part in its operations". As we will argue in the final chapter of this book, the discursive (and material) formation of the 'performative self' through these practices is central to the realisation of the new theory of human capital and is pivotal in the new modes of governance which emerged in the last decade or so.

Conclusion

This chapter has analysed the effects of globalisation on educational policy production within and beyond nation-states. This has entailed analysing the 'deconstruction' of the nation-state and its policy producing apparatus from both global political pressures and from associated fragmenting tendencies internal to it. Postnational politics work in relation to both tendencies. The globalisation of the economy has witnessed an uncoupling of highly mobile capital from the bounded and territorialised space of the nation as a site of policy production. This uncoupling has brought about structural changes in the modes of governance within the nation. It engendered a new consensus in education focussed on human capital which interprets education, in terms of both its quantity and quality, as necessary to the competitive advantage of nations. The apparent lack of coincidence between transnational flows of capital and the Keynesian bureaucratic state has precipitated new modes of governance in education. The result has been an erosion of public/private divisions and a range of new policy and ideas networks grouped around specific educational issues that accompany the procedure of steering at a distance. These networks work above the nation through the activities of international organisations such as the OECD, and in some instances at a supranational level in the case of the European Union.

In the context of globalisation, international (and supranational) organisations take on an enhanced policy role. They contribute towards an emergent global policy community, a point which returns us to the OECD. The OECD certainly sees itself as a key player in a globalising milieu. Indeed, the quotations opening this chapter came from the OECD report *Towards a New Global Age: Challenges and Opportunities* (OECD, 1996f), an explicit expression of the Organisation's support for economic globalisation and the liberalisation of world trade. The document highlighted the potential of globalisation to create losers as well as winners. It focussed, however, on strategies for realising the potential benefits. These were wide-ranging of course, but ultimately, the document stressed, the economic and social problems of member and non-member countries would have to be addressed through domestic policy reform, including education and training and "radically freer labour markets". The OECD cast itself in a key role of promoting this process of reform: "explaining to governments and the public that globalisation is a 'good thing' "; counteracting the fact "that the press tends to focus on the 'bad news' of globalisation, while rarely mentioning the 'good news' "; working more closely with politicians; helping governments to transmit policy messages to their own publics; and developing "a role as an educator in economic matters" (pp. 16–17).

The Organisation appears to be engaged in an ideological crusade, in which it simultaneously tries to steer the processes of change while being itself reshaped by those same forces. The same dynamic also reshapes its constituency — member countries. In these processes of change, and despite the apparent educational policy consensus across OECD member countries, there is disagreement over the ends of education and governance both within and without the Organisation. These matters are examined in more detail in the chapters that follow.

3

The OECD, Globalisation and Educational Policy Making: Changing Relationships

"Certainly such influence as the OECD has had on national developments derives not so much from the generation of new ideas by the Organisation itself as from its ability to pick up new ideas, from research and political stances in the countries ... develop their potential for implementation and then bring them to bear more broadly on national policy agendas. Free of executive responsibilities in this field, the OECD was all the more equipped to exercise this catalytic and integrative function without which a number of developments in the countries would at least have taken much longer to occur."

(Papadopoulos, 1994, p. 203)

Those concluding comments from Papadopoulos's book raise a number of questions about how the OECD operates and wields influence. The comments present a particular view of how the OECD approaches its work, based on assumptions of there being "an equality of thought and decision-making" (IT 56: secretariat, 1995) among member countries. Each has its own policy traditions but each is bound to the forums of the OECD by shared policy interests and concerns. Pressures from globalisation may require rethinking both the relationship between the OECD and its member (and non-member) countries and how the OECD influences educational policy making. These issues will be addressed in this chapter.

A useful starting point for this discussion is provided by Archer's (1994) more general framework for analysing the role and functions of international organisations. He argues that international organisations are implicated in policy making at three levels: as instruments of policy; as policy making arenas; and as policy actors in their own right. As policy instruments, they may be used to identify problems, inform national debates or legitimate already-taken policy decisions; they enable formal, diplomatic interaction between member states while also serving as a "battleground for individual members or groups vying for influence over and control of the organization" (p. 9). As policy arenas, they provide a "meeting place where members can discuss matters of common interest" and "a platform from which members can espouse views and ... confront each other" (p. 10). And as independent actors, they become an identifiable entity, "distinguishable from [their] member states" (p. 11). Archer cites the European Union and the European Court of Justice as examples of Organisational actors with decision-making capacities over and beyond their membership. The OECD's independence is of a different, more tenuous order, deriving in part from the secretariat's agenda-setting capacity and in part from the authority of 'the OECD imprimatur' — seen for example in 'its' authorship of publications overriding individual authorship.

When applied to the OECD, Archer's framework offers some analytical purchase for examining the way the Organisation works, in particular since it opens up discussion of potential tensions between different levels. Take, for example, this seemingly straightforward self-description:

> "[The OECD] offers governments the opportunity to debate, share experiences and find solutions to common problems arising in their national contexts. The Organisation is entirely at the service of its Member countries ... It is a forum for objective, skilled and independent dialogue which allows the broad understanding and in-depth comprehension required to deal with problems posed by an increasingly complex world."
>
> (OECD, 1998a, p. 2)

However, some critical questions should be posed in relation to this description. How does the relationship between the Organisation and its member countries work? How might it be changing? In an increasingly complex world, for whom does the Organisation act as a forum? What kinds of common problems and solutions are identified in OECD forums? Such questions may serve to unveil the deeper problematic of how globalisation may affect the OECD's relationships with member and non-member countries, its education agendas and its affiliation with the broader policy community of international organisations and agencies. To explore the issues in more detail, we will examine changes in three related aspects of the Organisation's mode of operating: the relationship with its constituencies; the ways by which it derives its education agendas; and its sphere of influence.

Who Are the OECD's Clients?

Formally, the OECD is an intergovernmental organisation and governments tend to see the OECD as 'theirs'. However, the Organisation's links to research communities and individual institutions as well as governments make it "more than simply the pawn of ministries" (IN 16: secretariat, 1995) and somewhat unique: "The OECD is a very interesting organisation because you have access to good researchers and to the thinking of national governments. Not many organisations provide this combination" (IN 2: secretariat, 1995). To some extent the OECD can be described as partially non-governmental in character, which poses questions about the nature of its constituency and the relationship between these two arenas, governmental and non-governmental.

Significant here is the role of the OECD secretariat, appointed on merit as 'international civil servants' rather than on the basis of country quota or representation and which, importantly, fulfils both administrative and research functions. At one level, the secretariat operates as the administrative arm of an organisation which is the instrument of its member countries. It prepares background papers and reports for committee meetings and conferences. It provides organisational continuity and memory in an agency where government members come and go. It generates the processes of consensus on which the Organisation relies. At another level, secretariat members constitute a professional body of researchers with some "freedom to manoeuvre" in terms of selection of project themes, organisation of meetings and choice of consultants for policy research and advice (IN 4: secretariat, 1995). How governments choose to use that advice is a moot point. Paradoxically, the OECD's influence and *cachet* stem to a high degree from its perceived independence from the particular interests of governments. As Alan Ruby, a former Chair of the Education Committee commented:

> "The multilateral nature of the organisation gives it credibility and authority as well as making it one step removed from individual governments. This gives it important degrees of independence which it can use to promote ideas which may be at variance with the policies of some governments."
>
> (Ruby, 1997, p. 12)

There is therefore an ongoing dynamic to be managed: of needing to both "draw outsiders in" and "push them out again" (IN 51, secretariat, 1998). The 'outsiders' are the consultants or experts who, for some, are the key to the Organisation's influence:

> "The OECD's strength is the consultants, not the small staff and the secretariat, that is groups of people who use the forum of the OECD to advance their ideas. ... What is interesting about the OECD are the networks of influence via the advisers and consultants. These are more important than the official connections, meetings, or publications."
>
> (IN 16: secretariat, 1995)

Perhaps wariness of these 'networks of influence' heightens governments' proprietary sensibilities:

> "There are some tensions between government views and the academics who may work for different programmes and who may well be resisting government policies."
>
> (IT 56: secretariat, 1995)

"The OECD can appoint experts on projects who may not necessarily be the one a government might appoint. But the outcomes must be for government use, not academics."

(IN 20: bureaucrat, 1995)

Not all secretariat members, however, view governments as the Organisation's unambiguous clients:

"Who are the OECD's clients? This is an interesting question. The clients used to be central governments, but increasingly this is problematic."

(IN 58: secretariat, 1995)

"The OECD deals with governments, nations and member institutions ... The notion of 'nations' allows for participation of provinces and small groups."

(IN 7: secretariat, 1995)

These statements draw attention to tensions inherent in the way the Organisation works — between its governmental and non-governmental character, and in its dealings with various types of federated nations. Over and above such tensions, however, the changing nature of the nation-state and the geo-political changes which are part of the process of globalisation may contribute to other changes in the relationship between the OECD and its constituencies. These we can comment upon from two perspectives: the country and, more latterly thematic, policy reviews which represent a significant component of the OECD's educational work; and shifts in what is often referred to as 'the OECD approach'.

Changing relationships: country and thematic reviews

The reconstitution of the nation-state has impact on the nature of the OECD's constituencies and on its mode of relating to them. The shifting dynamics are revealed in seemingly unremarkable ways: the break with the tradition of country policy reviews with the review of the higher education system of California in 1987, described at the time as "now one of the leading powers of the world" (OECD, 1987b, p. 3); the contribution to an 'Educating Cities' project in 1992 to underline the importance of the 'city factor' in educational policy making (OECD, 1993b, p. 2); the introduction in 1995 of thematic reviews "to give a horizontal rather than a vertical view" of key issues across countries "or units within countries" (IN 2: secretariat, 1995); and the increasing rate at which country reviews are being sought by would-be or non-member countries. Taken together, these developments point to a shifting mode of operation. In this connection it is worth analysing the country and thematic reviews further.

Country reviews constitute a major undertaking in the work of the Education Committee.

"Each review is intended to provide an overall assessment of the functioning of a country's education and training system but country authorities also identify the policy fields which they consider important and expert reviewers are chosen by the OECD to reflect these aspects. The focus of the reviews thus usually reflects both the preoccupations of the governments involved and the OECD's overarching concerns."

Townshend (1996)

While regarded as influential, country reviews are in many ways idiosyncratic, offering no formal mechanisms for comparison though, where possible, reviewers or 'examiners' (few of whom are women) are chosen who can bring a comparative perspective to bear on their reports (IN 2: secretariat, 1995). Informally, countries may extract a comparative benefit: "Conducted in an OECD context a review has the added value of informal comparisons by educational peers with experience of differing political and cultural background" (Kogan, 1979, p. 57). As Kogan points out, the reviews reflect similarities in context and a degree of 'parallelism' in policy responses over time. In essence, country reviews represent the quintessential 'OECD approach', captured by Kogan's comment: "Different countries have different expectations of the reviews. This is compatible with the OECD view, that there is no restriction on what might result" (p. 56).

Maurice Kogan, a long-time OECD consultant, examiner and editor of the IMHE Journal *Management in Higher Education*, was asked in 1979 to conduct an evaluation for the Education Committee of its fifteen-year span of reviews. His report noted that the idea of conducting thematic examinations over more than one country differed from the normal 'exchange of views' on themes of common interest. The proposal for thematic reviews, he observed "invites the OECD to select sharp and focussed issues arising from major current policy matters that countries in common need to settle rather than simply to discuss" (p. 71). This idea he firmly rejected: "First, it will reduce the national motivation. There can be a healthy narcism in the examination approach. Secondly, it might produce facile comparison" (p. 71). While he conceded the need for better collation of country reviews and ongoing thematic studies to strengthen the Organisation's policy work, his concluding comments reaffirmed his methodological and philosophical stance:

" . . .   a successful examination should seek to provide an analysis in such a form as to be directly useful to the country, and thus respect the immediate problems the country faces as being the main focus of the enquiry."

(p. 71)

" . . .   the background report and the examiners' response [should] continue to be rooted firmly in the experiences and problems of the country examined. National histories and educational policies should not be blended into generalised metahistory."

(p. 75)

To focus on country particularities would, he argued:

" . . .   put the OECD in a strong position to give leadership to the development of policy studies that are reflexive, because empathetic to perceived needs, and not derived from the rhetoric of externally imposed change, or from the narrow imperatives of economic analysis"

(p. 75).

Kogan's report showed strong support for the review process from member countries despite some criticisms. In recent years, however, this has changed. Reviews have been increasingly sought by non-member or candidate member countries, for example the Russian, the Slovak and Czech Republics (the latter a member since 1995) and Hungary (which joined in 1996). This was welcomed by the OECD as a means of "bringing non-member countries into the OECD's work on education' (OECD, 1992a, p. 80). At the same time, member countries have become somewhat reluctant to participate (IT 17: secretariat, 1997; IT 56: secretariat, 1995). In the early 1990s and in the face of initial resistance from the Education Committee, the idea of thematic reviews focussing on key policy issues of common concern to member countries was introduced as one way to 'bring countries back in' (IT 56, secretariat, 1995). In this, the Organisation was successful. The two thematic reviews conducted to date, on the first years of tertiary education and the transition from initial education to work, have produced "an embarrassment of riches" in terms of country support, pushing secretariat resources to their limit (IT 14: secretariat, 1996).

The significance of introducing thematic reviews should not be overstated: the country review process continues. In some ways it may have been strengthened by the introduction of a monitoring element. In any case there exists an inherently thematic element in the OECD's work. What is significant is what the shift may represent in terms of the approach for identifying policy problems.

From an OECD approach to an OECD position?

During the 1980s the relationship between education and the economy became a major priority for analysis as member countries started to grapple with the educational implications of a globalising economy. Along with this went changes in how the Education Committee and CERI Board approached the task of deriving their policy focus — changes which can be gleaned from the programmes of work submitted to Council and from the review of education (from 1993, education and training) activities contained in the Secretary-General's annual report. The report for 1980 states that, in devising their programmes of work, the Education Committee and CERI Board were:

"guided by the specific approach of OECD in relating educational problems to the changing social, economic and technological context in Member countries, and by the priorities which were defined in the Declaration by Ministers at the first meeting of the Education Committee at Ministerial level in October 1978."

(OECD, 1981, p. 58)

The report for 1981 noted that the renewal of the educational mandate for the period 1982–1985 was made:

" . . . in recognition by Member countries that the Organisation provides a unique forum for industrialised countries to pursue co-operation in education among themselves; and [in] recognition of the value which these countries attach to the approach which the OECD applies to its educational activities in relating them to policies in other sectors."

(OECD, 1982, p. 58)

Recognition of the concerns of member countries continued. But, the introduction of the new five-year educational mandate for the period 1992–1996, reveals a subtle but important shift in how the Organisation described itself. The report for 1991 records that the theme of "high quality education and training for all'' was selected "based on the medium term strategic objectives of the Organisation" as well as the outcome of the Ministerial meeting on education in 1991 (OECD, 1992a, p. 82). In other words, the Organisation — the actor — now receives equal billing with the member countries. Thereafter, in both the draft programmes of work of the Education Committee and CERI Board, and in the annual reports, there is little direct reference to the concerns of member countries. By definition, the policy agenda reflects the priorities of member countries. However, whereas this condition was always stated explicitly, after 1991 it is the *themes* selected for analysis which are highlighted.

The changes give an impression, first, that the policy agenda is driven as much by the external imperatives of context as by the internally derived concerns of member countries though the two arenas are of course related; second, that a more overt 'OECD position' is being set in place, as distinct from an 'OECD approach'. The latter term is used in a number of ways, both self-referentially by the Organisation as well as by outside observers. Istance (1996, p. 94) took the view that a significant element of the OECD approach to education is the principle "that national policies and practices should be informed through reference to those of other countries, learning from their achievements and failures but not blindly copying". That frame of reference was certainly never abandoned. Indeed, given the perception of the increasing commonality of problems, the propensity for countries to want to learn from each other's experiences, if anything, strengthened over the 1990s. At the same time, it is arguable that over the past decade the essential OECD approach in some respects has been overlain by a more normative stance. In 1991, the Education Committee asserted the importance of making "best possible use, in a context of rapid technological change, of our countries' prime asset, namely their human resources" against the backdrop of "the changing international context and the globalisation of policies" (OECD, 1991, p. 7). Such declarations could be seen at one level as espousing a rhetoric touching on concerns common to all of its member countries. At another level, however, they could be interpreted as an ideological articulation of educational purposes giving shape to the parameters of policy making everywhere, reflective of what the previous chapter described as the new global educational policy consensus.

While we would not want to overestimate what may simply be a shift in reporting style, nor read too much into the determining nature of such rhetoric, we would suggest that by the early to mid-nineties the OECD had indeed, in Kogan's terms, moved to "select" issues "that countries in common need to settle rather than simply to discuss" (1979, p. 71). Interesting in this regard are the changes in emphases in the Organisation's self-promotion. A publicity booklet, undated, but available prior to 1994, made brief reference to the Organisation's *modus operandi*: "Mutual examination by governments, multilateral surveillance and peer pressure to conform or reform are at the heart of OECD effectiveness" (OECD, undated, p. 10). A later edition is more prescriptive:

> "This 'multilateral surveillance' depends on the peer pressure system that has traditionally operated at OECD — examinations in which a government that fails to meet its commitments may be subjected to moral pressure by its partners. The effectiveness of policies increasingly requires mutual adaptation and, not infrequently, accepted guidelines that all governments may follow — or must follow, depending on the agreement — in formulating their policies ... "
>
> (OECD, 1994a, p. 9)

And more recently still:

> "This 'peer pressure' system encourages countries to be transparent, to accept explanations and justification, and to become self-critical. This encouragement for self-criticism among representatives of Member countries is the most original characteristic of the OECD."
>
> (OECD, 1998b, p. 2)

The notion of self-criticism is not novel. It was always implicit, and widely accepted, in the country review process. What is new is the elaboration of what is meant by 'peer pressure to conform or reform'. That idea always entailed a degree of question-begging which, in the context of 'the OECD approach', could be left open. By contrast, the notion of adaptation to 'accepted guidelines' hints at a subtle change of orientation in the Organisation's *modus operandi*: it moves from neutral forum for exchange of ideas through which processes of peer pressure may operate, to become a policy actor with its 'own position' against which policy stances might be judged.

Such a development in relation to the country review process had been raised earlier in Kogan's appraisal when he asked:

> "Is there legitimately a set of criteria for policies and the processes of policy-making against which individual countries can be judged? If so, how might those criteria be used sensitively, and be themselves capable of change ... "

Probably not, he seemed to conclude:

"There seems to have been a common climate for educational policy, but common conditions need not justify common criteria because educational development is most secure when it grows from within … "

(Kogan, 1979, p. 10)

Given these reservations, the proposal to develop monitoring procedures for country reviews casts an interesting light on the Organisation's tentative moves down a more judgmental track. To the three stage country review process (preparation of a background report; the examiners' report; the examination or 'confrontation' meeting plus published review), a fourth stage was added in 1996 which involved "a further review of performance by the committee two years later, to examine whether recommendations have been implemented" (OECD, 1996c, p. 14). In effect, the review process is still evolving. It remains non prescriptive, so that "countries can — and do — indicate why they have chosen to go a different way" (personal correspondence, OECD). Nevertheless, monitoring adds an extra dimension to the country review process, a dimension which interestingly has come into play at a time when many new countries are seeking to join the Organisation.

Defining the OECD's constituency has never been straightforward given its dual relationship with governments and research communities or institutions. But the nature of the relationship has evolved. This is to be seen in changes to the process of country and thematic review and through shifts in the Organisation's *modus operandi*. These changes, we suggest, reflect the shifting geo-political terrain which is a by-product in particular of political globalisation. Globalisation processes are also having an impact upon the Organisation's policy priorities and upon its choice of specific project themes — in short, upon its educational agendas.

Shifting Agendas

Agenda setting and the secretariat

Agendas emerge from a complex mix of factors, including "accidents of time and person" (Kogan, 1979, p. 65), happenstance and persistence:

"There's a conference coming up, or world summit on something, and then suddenly there's a request that you should look at globalisation and women and education for example … the Beijing summit on women for example. So in the run-up for that there comes a request through various ways for us to contribute a paper on globalisation and the economy."

(IT 17: secretariat, 1997)

"Things happen by chance quite often … An idea comes up at Education Committee, and [head of Division] asks: is anyone interested in this one?"

(IN 51: secretariat, 1998)

" . . . the reason for that [particular idea persisting] is simply that there is a group of countries that is totally committed and they don't give up, they don't want to buy no for an answer, and . . . they continue to provide resources, their own resources, to pursue the objective . . . as long as we provide a forum for those countries."

(IT 17: secretariat, 1997)

Behind these sometimes irrational or erratic circumstances that invite policy attention are more systematic factors, usefully described by one member of the secretariat as: the constellation of staff in the secretariat (and to which the consultants may be added); the political and economic climate of the times; and the internal dynamics of the Organisation (IN 2: secretariat, 1995).

The OECD's merit-based recruitment system delivers a professional and relatively independent secretariat, characteristics which contribute substantially to the quality of the reports and analyses on which the OECD's reputation rests. Over and beyond that, there is a sense in which the secretariat is seen as "at times very strong, at other times weaker" (IN 2: secretariat, 1995) or as more or less 'active'. These terms are value-laden, but in essence refer to the capacity to "confront the economists" (IN 15: secretariat, 1995) and set an independent agenda for education.

"There was a very strong secretariat in the period 1964 to 1968. ELSA [Employment, Labour and Social Affairs] then set the agenda on education. Problems arise when people are not strong enough to confront the economic analyses."

(IN 2: secretariat, 1995)

The impression conveyed by Papadopoulos (1994) is one of a peak period of secretariat activism in the early 1970s, assisted by the recruitment of some former student activists of the 1960s, a positive economic climate, widespread support for Keynesian economic policy and accompanying political optimism. Indeed, Kogan (1995), countering popular perceptions of the Organisation as an agent of Western imperialism, pointed out that OECD educational reports were often criticised for their social democratic bias. Perhaps such views reflected the reality that "most educational policy making until the 1980s was indeed 'social democratic' even if the parties in power were of the right". To come to the second factor, agendas are also a product of the political and economic climate of the times, "expressed through the stamp imposed by the governments, reflected in recruitment patterns and degree of openness in the organisation" (IN 2: secretariat, 1995). So for example:

" . . . in the 1970s, the combination of a strong secretariat plus strong government interest provided the impetus for expansion, the establishment of the various programmes and committees in education, and it affected recruitment. During this period, there was a focus on equity, justice etc, and these themes had an echo in the member countries, and their support."

(IN 2: secretariat, 1995)

Arguably nowadays the secretariat is more attuned to the thinking of economists. In part, this receptiveness may well reflect recruitment patterns within education which favour the appointment of economists rather than, say, sociologists or philosophers (IN 50: secretariat, 1995); in part, it may also be an outcome of the increasing emphasis on 'horizontality' within the Organisation which helps to integrate economic and educational analyses more closely; and, in part, the orientation may reflect the predominant government preoccupation with economic imperatives, exerting "a subtle chemistry" on the way education works (IN 12: secretariat, 1995). Although the social purposes of education and matters of equity have not been neglected, they have in some ways been residualised. Only very recently, against a growing clamour over the socially unacceptable consequences of globalisation, have conditions emerged which might make for a different kind of prioritising:

> "Now we have the chance to do it again ... we mustn't let current concerns with social problems, which are recognised by the economists, be taken over by them."
>
> (IN 2: secretariat, 1995)

To the secretariat should be added the consultants, a relatively closed network of people with whom secretariat members build up relationships over time. What qualities are sought? First, "people we know" (IN 12: secretariat, 1995):

> "Once consultants are chosen by the OECD, they tend to be re-employed over and over."
>
> (IN 44: bureaucrat, 1996)

> "And that meant my name was on somebody's desk when this other task came up ... I don't have a strong background [in that area] I must say, but [x] knew me enough I guess to have some confidence that I could do this other job."
>
> (IT 14: secretariat, 1996)

Secondly, people who are known to deliver on time:

> " ... you are working to very tight deadlines and you go for people who you know can deliver directly from your own experience, or that you know that other people have had success with in the past — So, informally, [these people] are now in the pool."
>
> (IT 14: secretariat, 1996)

Thirdly, people who fit the OECD 'norm'.

> "There's an element of safety in who gets appointed. For example, [x, known as 'radical'] would be an unlikely choice, he doesn't have a sufficient degree of flexibility. You need to be able to take account of the fact that others have different views."
>
> (IN 35: consultant, 1996)

"Some things are not on the agenda. For example, [x] was against increasing school retention, which meant he was written off as a consultant ... extreme ideas, left and right, are written off. You can't afford to be too far outside the mainstream or you'll be ignored."

(IN 31: consultant, 1996)

In large measure, the consultant pool reflects the proclivities of the secretariat. Attention now turns to the third factor, to the internal dynamics of the Organisation, which concerns the relationship between the powerful economic directorate and the rest of the Organisation:

"The 'E' in OECD is always economics, not education ... These [economic] pressures came from within the organisation, not from outside. They then set the context for how educational issues are taken up."

(IN 2: secretariat, 1995)

"Power lies with economic directorate, particularly finance. Education works within the parameters set there."

(IN 18: secretariat, 1995)

"Papadopoulos's book gives an account of education from an education perspective. There is another perspective: that of council, where education is hardly discussed. Its emphasis is finance and it is much more prepared to watch the economic codes. Economics takes up about a third of council's time — the other directorates get much less attention."

(IN 49: secretariat, 1997)

In the past, some argue, the relative marginality of education in the broader scheme of things provided an independent space for educational work (IN 2 and IN 6: secretariat, 1995). Now, however, education has found its place in the economic sun; it was, for instance, the focus of specific attention in Australia's 1997 annual *economic* review for the OECD. DEELSA was pleased to contribute to this work, seeing it "as bringing some of its substantive ongoing work into central policy focus, because the national policy makers in Australia take a lot of notice of the annual economic survey" (IT 14: secretariat, 1996). The tendency to meld education and economic policy reflects the loss of education's *sui generis* character.

Evolving policy priorities

The wide range of themes and topics addressed over the years by the OECD's four education programmes has already been noted. The evolving policy priorities underpinning these various projects and activities will be examined against a background of the growing preoccupation with 'the changing international context and the globalisation of policies' which has already been touched upon.

First, the evolving policy priorities, beginning with the renewal of the educational mandate for the period 1982–1986 will be briefly sketched out. This mandate was informed by the priorities coming out of the Ministerial conference in 1978 on Education and Economic Growth. Six areas were designated under which activities were organised during that time: interaction between education and policies in other sectors in changing social, economic and technological conditions; the contribution of education to improving the preparation of people for working life; educational provision for disadvantaged groups; changes in compulsory schooling; access to, and reorganisation of, post-secondary education; facilitating the exchange among member countries in education innovation (OECD, 1982, p. 59).

These priorities were clarified further the following year:

> "The central issue ... is how, in a situation of cut-backs in public resources, the dynamism of the educational sector can be maintained so as to satisfy new social demands brought on by deep structural changes in the economy and society. Member countries are increasingly dependent on highly skilled and adaptable populations capable of promoting technological change, industrial development and social progress. Education and training have to respond to this challenge at a time when young people face the prospect of unemployment."
>
> (OECD, 1983a, p. 61)

Structural change, unemployment and skills formation were to form a *leitmotif*, reflected in the distillation of priority areas to just two when the next mandate was introduced for the period 1987–1991. The two areas were: "the contribution of education to social and economic restructuring; and reinforcing the quality of education" (OECD, 1987a, p. 55). In 1989, concern lay with "the over-riding importance of education and training for the efficiency of the economy and the ability of society to change" (OECD, 1989a, p. 1). The two priority areas were reorganised as the Organisation's medium-term priorities for 1990–1991, namely: employability of individuals; access to education and training; and quality of education and training. The year following they were in turn brought within an overarching framework of five key concerns, drawn from the 1990 Ministerial meeting on 'High Quality Education and Training For All'. They were relevant education and training for all; strategies for life-long learning; implications of the international dimension; new partnerships in decision-making and resources; improving the knowledge base and communication (OECD, 1992a, p. 82). [2]

Priorities for the five-year mandate 1992–1996, were clearly influenced by the policy orientations flowing from that Ministerial Meeting as well as the Directorate's new medium term priorities: human resource development; monitoring

---

[2] The reference dates in this section may be confusing. At times we draw on the Secretary-General's annual report which covers activities in the year preceding publication. At other times we draw upon the Education Committee and CERI Board's draft programmes of work which refer to activities to be conducted in the following year.

of labour markets and social change; and the global policy dimension (OECD, 1991, p. 7). Within this framework, activities for 1992 and 1993 emphasised three major themes: links between education and labour markets; improved ways for measuring and assessing educational processes and outputs; and initial collaboration with non-member countries in Central and Eastern Europe (OECD, 1993a, p. 79). Apart from the ongoing preoccupation with education–economy links, these themes reflected the Organisation's expanding work in non-member states and countries and a heightened emphasis on cross-Directorate cooperation to carry out horizontal analyses of the relationships between education, social and labour market policies (OECD, 1995a, p. 83), for example the Organisation-wide Jobs Study Project on employment and unemployment which commenced in 1993. Also reflected in this framing is the increasing prominence of work on educational indicators, with the first edition of *Education At A Glance* appearing in 1992.

The themes emerging over this time were summed up in the Directorate's 1994 medium-term objectives which required that 'special attention' be paid to (a) globalisation, (b) growth, employment, social problems, and (c) openness to the rest of the world (OECD, 1994e, p. 5). These objectives set the frame around three themes: a coherent, comprehensive framework for the development of human resources; monitoring and review of labour market, education and social policy developments; and the global dimension (*ibid.*, pp. 5–8). The priority of considering "alternative policy directions for making *lifelong learning* a strategy for overcoming social exclusion and responding to continuing structural changes in national economies" evolved from this background (OECD, 1996a, p. 86, original italics). Thus the theme of the most recent education mandate 1997–2001, 'Lifelong Learning For All' shapes the present 'operational objectives' (formerly called areas of interest) of the Education Committee and CERI Board: monitoring and evaluating policies in an international setting; strengthening the foundations for lifelong learning; facilitating transitions through lifelong learning and work; mobilising resources; rethinking roles and responsibilities of governments and partners (OECD, 1996b, p. 4).

So what is the picture that emerges? At one level, a patchwork of loosely connected topics (and even more so the sub-themes which were not examined) have been strung together to provide a minimum semblance of coherence. In particular, the general perspective of education's contribution to social and economic restructuring seems to provide the glue by which priorities have been attached — employability of individuals, access to education and training, development of human resources, facilitating transitions through lifelong learning and work, relationships between education and labour market policy. Three other important strands emerge: the international or global policy dimension; rethinking roles and responsibilities of governments and partners; and monitoring and assessment. From one standpoint, with the exception of the explicit mention of globalisation and internationalisation, these are not new themes. But the ways in which these themes became reworked over the late 1980s and 1990s reflect a new perspective in educational policy and governance which, we would argue, relate directly to pressures exerted by globalisation.

From a different standpoint, arguably, the unfolding policy agenda does indeed reveal a more coherent story about the way the OECD has been both an actor in globalisation and target of globalisation pressures. In particular, the OECD's articulation of the new micro version of human capital theory described earlier, and its advocacy of new accountability mechanisms for emergent global education systems and markets, point to the OECD's role as a key exponent of the new policy consensus, a consensus grounded in developing skills for a global, knowledge-based economy and new models of educational governance. In this, however, the OECD is hardly alone. Rather, it has become enfolded into an emergent global policy community, which has implications for how the Organisation operates and wields influence.

Changing Role and Sphere of Influence: From Think-Tank to Policy Actor

The 'new world order' is exerting pressures on the OECD. New regional groupings and the formation of the EU have resulted in some loss of adherence from traditional members. At the same time, many would-be member-nations which now meet the formal qualifications of economic liberalism and democratic pluralism seek to join the OECD club and gain the "OECD imprimatur" (IT 56: secretariat, 1995). Two related issues follow from this changed configuration: greater pressure of work; and the need to redefine and identify a distinctive and specific role for the Organisation.

Work pressures are evident both in terms of the increasing demands made of the Organisation, especially from non-member countries of the former Soviet Union and Eastern Europe, Central America and South East Asia, and the speed at which it has to respond to them. The situation is made worse by budget cuts of 10% — largely at the insistence of the US and with Australian backing — with the resultant loss of 200 posts between 1997 and 1998 alone. Loss of staff and rising demands have brought about, some argue, a shift in the way the OECD functions, with the potential for its distinctive think-tank *niche* to be replaced with a somewhat different orientation, sometimes described as "becoming more practical" (IN 20: bureaucrat 1995), as entering the "real world", as facing problems "it hasn't really had to confront before" (IN 23: international organisation, 1995), or being used by governments to provide "quick and dirty solutions" rather than for solving problems (IN 15: secretariat, 1995).

Critics in the Organisation argue that the intensification of work has undermined its capacity for reflection and research, and diminished the Organisation's prime analytical capacity. It is an environment of "quickie conferences and meetings, badly planned and prepared, without proper briefing papers and which nobody reads anyway" (IN 6: secretariat, 1995), and of declining resources, many extra requests, inadequate administrative assistance, late reports, and no proper discussion process (IN 2: secretariat, 1995). It is an atmosphere of being:

" ... caught up in management pressures to perform, and quickly – 'how many days to do a [project]' sort of thing ... Of course, it's much less efficient, because doing things in a hurry involves lots of telephoning because the good people aren't available to write something up in a hurry. So you spend more time on the phone and in addition you're left with a mediocre person to do the job."

(IN 50: secretariat, 1995)

Not all see it that way of course. For some, the new environment serves to sharpen the OECD's distinctiveness, specially vis à vis the European Union:

"It's convenient for countries to have an organisation which doesn't have the pre-scriptive or the financial powers that the Europe Union has but [which] must always operate on the principle of a kind of equality of thought and decision making, peer review, and those kinds of modes of operation."

(IT 56: secretariat, 1995)

In the words of another interviewee, although the pull of the EU was strong, "countries are still very keen to be here because they know that on the conceptual, analytical level the OECD is still far ahead" (IN 49: secretariat, 1997).

How is the OECD seeking to redefine its position? Underpinning this discussion is the question of the OECD's 'three faces' described earlier in this chapter — as policy instrument, forum and actor — and their changing character. The dilemma is neatly caught in this comment:

"There is more propensity now for governments to want to think through ideas than before, given the current pressures. The secretariat itself has to rethink things. An aspect of this can be seen in links with the NWTO (New World Trade Organisation) ... there are complementary roles. That is, ideas are hammered out at the OECD and then negotiated at NWTO. It needs good thinking for this. If we can't keep that function, there is no rationale for us."

(IN 2: secretariat, 1995)

This statement poses some key issues: the need to rethink; the kinds of ideas which need to be thought about (links between education and trade); the nature of the relationship between the OECD and other international organisations. The statement highlights the increasing weight of the OECD as a policy actor as it begins to position itself within a global policy community of international organisations and supranational bodies which is also in process of definition.

The OECD, as argued in Chapter 1, is part of a developing network of international organisations which together may eventually serve as world level agencies in educational policy convergence and change (McNeely and Cha, 1994). They argue that international organisations influence the incorporation and diffusion of educational ideologies and practices within and among nation-states. International organisations, so McNeely and Cha assert, "have been an important catalyst in spreading world cultural themes and accounts". The presence of a representative from the European Commission on the governing Council and other forums of the OECD is symbolic of this trend, so too, is the tendency of former OECD staffers to take consultancies with the EU and other tendering agencies in Europe. The ease with which ideas now circulate amongst these groups via internet and email, reflect Appadurai's (1996) more general point that globalisation, modern communications and technology close the distance between policy elites.

While it is important not to run the policy stances of these organisations together, the congruent rhetoric on educational issues is remarkable, especially around such motifs as 'lifelong learning for all', the recurring rhetoric of quality, diversity, flexibility, accountability and equity. Such congruence may reflect a conscious effort "to harmonize the activities of international organisations so far as is practicable" (OECD, 1992b, p. 1). In some areas, these organisations could be viewed as an emergent global policy community constituted by an overlapping membership of senior public servants, policy makers and advisers. They form part of the group to which Sklair refers as "globalising bureaucrats", individuals who are:

> " ... active in powerful international organisations, notably the world bank, IMF and the OECD, and they also work politically through what have been termed 'corporatist' agencies that combine representatives of the state, business and labour."
>
> (Sklair, 1996, p. 5)

Sklair argues that the culture of this group, and the ideology it promotes, are a mixture of neo-liberalism and global nationalism "the view that the best interests of the country lie in its rapid integration with the global capitalist system while maintaining its national identity ... " (p. 5).

Of course, not all would agree with this assessment of what an emergent global policy community might look like. Suffice to say that the new world order presents the OECD with dilemmas in terms of its membership and in terms of carving out its own terrain. There seems little doubt that this geopolitical volatility obliges the OECD to look outwards more than it has — towards the Pacific, Asia, Eastern Europe and Latin America — seen in its efforts to "regionalise" itself and to hold meetings in the world beyond Paris (IT 56: secretariat, 1995; 13: secretariat, 1995; 53: secretariat, 1997). But this dynamic has consequences for the Organisation's distinctive character:

> "We are now twenty-nine ... Korea is the last one. We have got in three Eastern European countries and then Mexico. You will end up with a totally different thing. At the same time you can't ... [keep out] countries like China, Malaysia, Brazil, Argentina, where all the growth in the world economy is today. So it's a dilemma, you see."
>
> (IN 2: secretariat, 1995)

> "How do you resolve that dilemma? At present, by moving very slowly. We have only one new member from eastern Europe, and one from the New World — Mexico, pushed in by the Americans because of NAFTA. Korea will come in slowly. So it

will be a slow and steady increase with increasing relationships with non-member countries."

(IT 56: secretariat, 1995)

As a global policy actor then, the OECD may exert both more and less influence in education. On the one hand, its mission is strengthened by the tendency towards convergence — more policy actors are enunciating similar messages albeit from differing positions. For the new or would be members from the former Soviet Union, for example, such messages may indeed constitute a policy "haven" (Kogan, 1995). On the other hand, with such convergence the OECD's message also becomes more didactic, more ideological. It becomes less attuned to the nuances of national or sub-national circumstances, with a possible loss, as Istance (1996, p. 96) puts it, of "the distinctive form of OECD intellectual 'value added' ". Here then is a shift in the policy-making relationship between the OECD, as an international organisation and think tank, and its member and non-member countries. This shift raises broader questions about the relationship between differing arenas or levels of policy making.

Global, International and National Levels of Policy Making: Changing Relationships

One way of viewing the relationship between the national, the international and the global levels of policy making is to see globalisation as a context for a relatively autonomous nation-state, with international organisations serving as instruments of governments or as forums for the exchange of ideas between governments. Fowler (1995) suggests that, while nation-states are influenced by global forces, these are filtered through the prism of national characteristics such as economic resources, policy-making processes, and national values. This is very much the OECD's stated approach. In Papadopoulos's (1994, p. 13) words:

"A basic precept in the whole approach is recognition of the fact that education policies are par excellence national policies reflecting the particular circumstances, traditions and cultures of individual countries. The notion of an international educa-tion policy — even if such a thing existed, which has not been the case at least so far — is altogether foreign to this concept."

From this perspective, countries engage in problems which, in light of globalisation pressures, may be driven by increasingly common imperatives. Viewed in this way, the OECD in its role as an international think-tank operates, and exerts influence, essentially as a comparative forum, accommodating both a sense of national autonomy as well as a sense of commonality among the like-minded. The process of comparison may also contribute to policy borrowing or adaptation across the boundaries of nation-states, leading to "universalising tendencies in educational reform" (Halpin, 1994, p. 204; see also Finegold et al., 1993).

There is another way, however, of thinking about the genesis of universalising tendencies. Referring to the 'strengthened international dimension' in the knowledge base needed for policy making, Hirsch (1996, p. 31) pointed out that:

"Whereas policy makers in the past may have drawn from international experience sporadically, in future they may have to do so more as a matter of routine. This development raises the question of how to strengthen regular channels for disseminating all forms of educational knowledge across international frontiers. The role of 'Institutional International Mediators' of knowledge like CERI [within the OECD] itself becomes interesting to evaluate in this context."

To conceive international organisations such as the OECD as 'international mediators of knowledge' predicates a different relationship between the national, the international and the global. Mediating knowledge carries different connotations from policy borrowing. Once conceptualised as 'knowledge mediators', international organisations may be seen as engaged in a process of 'brokering' meanings — a process in which national distinctiveness may give way to more universal/global meanings negotiated via a network of influential international organisations. Some sense of what is involved may be illustrated by referring to the OECD's influential work on comparative educational indicators, published annually in the series *Education at a Glance*. While at one level the indicators project is about comparison of national education systems, at another, normative level, it brings into play what could pass for a global politics of comparison which has to do with supranational forms of agenda setting. Policy convergence may come about not simply as a result of policy borrowing in the face of common problems, but rather by the ways in which the bounds of education policy making within nation-states, *and* within the international forums where nation-states participate, are shaped by global forces transcending the frontiers of nation-states.

Bearing this in mind, it could be argued that, by its thematic reviews and its comparative work, the OECD acts as an international mediator of knowledge rather than as a comparative forum alone. Viewed thus, it is arguable that OECD policy agendas, as much as the processes of comparison themselves on which its reputation stands, serve to establish a universal (Western) norm against which the policy values of individual countries are, in the expression of Robertson (1992), relativised. This viewpoint was developed by some members of other international organisations who were critical of the tendency for OECD norms to permeate developing countries as universal norms:

"There are large differences between OECD countries, especially in economics, so I wonder how they ever do reach consensus ...   But there are elements of ideas [for example on public sector reform] which do nevertheless become adopted as the OECD norm against which all countries are compared and some are seen as 'leaders' and other seen as lagging behind ... "

(IN 22: international organisation, 1995)

"The OECD is very influential in this regard. The World Bank then picks up OECD ideas, and these get filtered into the underdeveloped world."

(IN 22: international organisation, 1995)

Ironies are not lacking. At one level, the OECD is perceived as having a more distinctive role: a standard-bearer of universal Western norms no less! But this elevation came at some cost to its other claim to distinction: as a think-tank or as a catalyst for change. The OECD juggles for position amidst the "international education mafia" (IN 25: international organisation, 1996). It is doing so within the narrow, albeit perplexing parameters of the moment, a quest not wholly devoid of disillusion:

"There's no think-tank now. That's all gone. The Education Committee is just the country reviews and the thematic reviews, and CERI is just the indicators. There's nothing much else."

(IN 51: secretariat, 1998)

While such a comment by no means applies (nor was meant to apply) to all of the OECD's educational work, it does echo the critique that educational agendas are being 'hollowed out' by the broader culture of performativity which Lyotard identified and which was described in the previous chapter. Such a critique ties in with the view of Muetzelfeldt (1995, p. 44) of the draining effects of market rationality which, he argues, lie at the heart of globalisation:

"On the one hand market like rationality and its instrumental approach to people and to social life has become more global, providing the dominant discourse and organising the dominant public practices of even larger areas of life across more and more of the world. On the other hand, the capacity of this rationality to provide political as well as social meaning and identity ... has been hollowed out. We are left with an overarching shell of abstract instrumentalism within which public identity, social institutions and everyday practices are increasingly drained of normative and communal content."

(cited in Smyth and Shacklock, 1998, pp. 16–17)

Such a conclusion may seem unduly pessimistic and too extreme. It is certainly very different from the pluralist assumptions which opened this chapter. It also lays bare the dimensions of power behind policy making and for which Lukes's second and third dimensions of power — controlling policy preferences and establishing frames of reference respectively — are apposite. Dale (1999) argues that the OECD exerts influence through the second dimension of power, effectively its agenda setting capacity. This capacity we hold to derive from the Organisation's modes of operation as a policy forum (or think-tank) and as policy actor. Yet, the "most effective and insidious" third dimension of power Lukes argues is the capacity to achieve consensus by shaping "perceptions, cognitions and preferences in such a way that [people] accept their role in the existing order of things" (Lukes, 1983, p. 24). Such an interpretation, though proceeding from a very different theoretical starting point, is akin to Foucault's knowledge/power nexus and to the Foucauldian notion of power as embedded discursively to govern "what can be thought and said" at any given time. One of the Leitmotifs developed in this analysis is that globalisation as a discourse (and as a set of material practices) operates powerfully to delimit what can be thought and said. Its implications for the way the OECD operates are no less paradoxical since the Organisation is both a globalising agent as well as being shaped in its turn by globalisation. Our view is that the OECD, as both the subject and the object of globalisation, is discursively bound by the parameters of its own globalising mission, and is hence unable to envisage radical new agendas. At best then — and in this both Kogan (1995) and Istance (1996) support Papadopoulos's contention that this is a lesser evil — the OECD can filter and synthesise existing ideas, and act as a catalyst. At worst, however, it may become an unreflexive carrier of neo-liberal ideology. The implications of this essential ambiguity beneath the Organisation's educational work are examined further in Chapter 4.

To summarise now what we have been trying to argue in these three initial chapters. The OECD was born out of the new international order established under the Marshall Plan following the Second World War. Economic reconstruction for a peaceable world order was predicated on principles of economic cooperation and interdependence alongside the Keynesian project of economic nationalism in which the state held a central role in nation-building, modifying the excesses of capitalism and generating some measure of equality of opportunity. The OECD and its predecessor OEEC were key articulators of, and a forum for, this post-war covenant which sought to conjoin national and international economic and social interests — a vision reflected in its charter of promoting a market economy and pluralistic democracy. This background also provided the impetus for what has sometimes been called the Keynesian policy settlement in education, that views education as a form of human capital investment with significant though immeasurable — i.e. residual — benefits for economic growth and social cohesion. On such pillars, the *sui generis* character of education rested and was reflected in the establishment of strong education bureaucracies in post-war nations and in the independent (albeit always interconnected) presence claimed for education within the OECD. As economic prosperity indeed became reality in OECD countries, this presence was affirmed by the establishment of CERI and by its wide-ranging body of research on educational reform, on innovation and by its support for the equality of educational opportunity.

Globalisation has overset that ideological and structural context. Economic nationalism has been superseded by the neo-liberal commitment to economic globalisation and by a smaller role for the state, fostering rather than mediating, market relations. In this new world order, the OECD is positioned between new regional blocs and agencies, the supranational EU, between more permeable nation-states and shifting constituencies. A post-Keynesian education policy consensus in education has emerged — vigorously espoused by the OECD — which is engaged in promoting both a reworked version of human capital theory and more distant forms of educational governance. Today, education is seen not as a residual, but as a direct and measurable factor in 'growing' the new knowledge-based economy which confers both individual and societal benefits and concomitant expectations. From this it follows, therefore, that individuals should bear a substantial proportion of the cost of their own education. Educational governance is no longer held to the business of state-funded centralised educational bureaucracies alone. Rather it is a partnership between a number of stakeholders. The upshot of the new policy consensus has fundamentally altered both the nature and the understanding of educational purposes. It has significantly changed the parameters of educational policy making. Educational goals have increasingly become merged and confounded with economic objectives on the one hand and, more recently, with broader social objectives on the other. In the midst of these shifts, the OECD as a "global intergovernmental economic organisation" (OECD, 1998a, p. 5) has been both an actor and acted upon, an ambivalence seen in its changing policy agendas, in its changing relationships with its constituencies and in its changing spheres of influence.

4

Ideological Tensions in the OECD's Educational Work

"OECD's underlying concern is . . . to take account of 'the numerous social problems besetting our countries, problems which, if we are not careful, could in the long run jeopardise economic development and even political and social cohesion.' "

(OECD, 1991, p. 7)

"It would be dangerous if this vague but widespread feeling of unease were to cause the very foundations of our countries' prosperity, namely the market economy, private enterprise and free trade, to be called into question. It is incumbent on the public authorities to appraise this danger and identify ways of guarding against it."

(OECD, 1992d, p. 5)

"OECD economies place "a premium on the development and implementation of coherent and effective policies for human resource development and life-long learning, bearing in mind the need to reconcile the efficiency and equity objectives."

(OECD, 1994b, p. 5)

Taken from the Education Committee's draft programmes of work in the 1990s these extracts encapsulate some of the abiding ideological tensions in the OECD's educational work as it dwells on economic and social demands and on their associated goals of efficiency and equity. Economic and social agendas surface in different ways within the Organisation, play off against each other in different ways, and are articulated differently depending upon time and context. Expressed differently, conceptualisations of desired educational purpose, and the means of arriving at it, have always been contested within the Organisation — between secretariats and within the educational research and policy forums. Thus tensions existed, for example, between what Deacon (with Hulse and Stubbs, 1997) calls the "IMF flavoured" economics secretariat and the social orientation of DEELSA, tensions which flow over into DEELSA's education activities. Over time, there have been periods when it was easier to pursue social objectives within and through education, and periods when economic imperatives prevailed. Overall, however, and within the overarching framework of the OECD's prime economic mandate, the education section sought to temper the view of education as handmaiden to the economy.

This view of the OECD's educational work contrasts with some of the more popular accounts which qualify it as simply 'neo liberal'. Take, for example, Smyth and Shacklock's citing with approval a description of the OECD as the World Bank's accomplice, wielding a 'Tarzan style' influence over education policy making:

> "A group of hairy-chested individuals swing down from the trees, uttering cries of 'efficiency', 'competition', 'market discipline'; they tip all the huts over, then they swing back into the trees, leaving the villagers to clean up the banana peels."

> (Smyth and Shacklock, 1998, p. 61)

Similarly, Spring's analysis of the OECD, while more nuanced, makes little mention of the deep disagreements which in our opinion typified the Organisation's approach to education. Spring's view is:

> "Education plays a dual role in OECD plans. First, education is to aid the development of market economies through human resource development and lifelong learning. Second, education is to remedy problems resulting from globalisation such as unemployment, increasing economic inequality, and fears of social and economic change."

> (Spring, 1998, p. 160)

The assumptions beneath such portrayals of the Organisation as a homogeneous unit with a narrow, static agenda, fail to capture its educational reach, the conflicts within its various forums and between the ideological layers within its charter which aim at both economic *and* social development and, more recently, respect for human rights. Our task here is to unravel the social and economic threads running through the Organisation's educational work. In so doing, we will try to eschew the simplistic dichotomies of 'economic' and 'social' — as indeed does the Organisation itself. Nevertheless, educational work within the OECD cannot ultimately escape the market liberal ideology which the Organisation propounds. In this respect we would agree with critics that the past decade has witnessed the ideological ascendancy of a particular view of the social purposes of education, a view that has resulted in such purposes becoming in essence residualised or embraced with some ambiguity. It has not always been the case and indeed this interpretation is still contested within the Organisation. It is a mistake to regard the OECD as upholding an ideological position, uniformly and monolithically neo-liberal, in educational matters.

The OECD's genteel and consensual style of working often masks the diversity of views on education within the Organisation, and camouflages the forms of confrontation over ideas and meaning that are tolerated, even encouraged:

> "There is no OECD line of education. They have projects running that have very different conflicting views. There's no one model that fits in with a view of the economy that they're trying to peddle around the world. This simply isn't the case, but there is a mythology that there is of course."

(IN 28: consultant, 1997)

The various positions the OECD assumed over the years are products of context as well as the internal politics of the Organisation. These are played out within the secretariat, between the secretariat and its governing committees and between member countries which, although predominantly European, reveal an ideological cleavage between social-democratic and neo-liberal stances on policy. The ideological divide is referred to within OECD parlance as 'European' versus 'Anglo-Saxon' though these descriptors hardly convey (nor are they really meant to convey) the variations within the two camps. Nor do they embrace the heresies which sometimes break out when individual representatives do not subscribe to their country's official position (IN 33: consultant, 1996). While much of the internal politics remains hidden away by the essentially consensual process of decision-making and extensive report-writing described earlier, it is possible to gauge how social and economic concerns have been expressed from time to time by analysing the Organisation's publications, its programmes of work, interviews conducted, and Papadopoulos's (1994) historical account which supplied an initial chronology of approaches.

Education has been broadly justified an OECD activity on the basis of its contribution to economic growth, though such a role is not necessarily interpreted narrowly. Papadopoulos, for example, marshals some persuasive evidence to argue that education was never looked upon within the Organisation "as a mere instrument of economic policy" (p. 12). He showed how early work in economics of education generated broader debate about the social purposes of education, which was the theme of the first major OECD conference on education in Sweden 1961, where "eminent sociologists presented detailed analyses of the major barriers to the use of potential human ability in education, specifically in terms of social class, rurality, school organisation and cultural inequalities . . . ". The Conference report, he argued, "remains a milestone in the sociology of education" (p. 33). He pointed to the large body of OECD work on educational disadvantage in the 1970s, which revealed a subtle understanding of the limitations of educational reform when dealing with essentially political questions about desirable social systems (p. 111).

Papadopoulos identified two broad eras in educational work: pre and post the 1970s recession, which parallelled the Keynesian and post-Keynesian periods of policy settlement mentioned in Chapter 2. The former, the 'golden era of growth', saw the establishment of CERI and CERI's initial work on educational innovation and reform, then defined in somewhat different terms from today. At that time, this body of work was guided by policy principles of economic nationalism. It was impelled by motives of expansion, of educational provision and equality of educational opportunity for nation-building purposes in a context where education was held to serve broader purposes than simple meritocratic selection into jobs:

> " . . . the role of education in equality will be properly stated when the educational system is used to develop a wider range of talents, abilities and attitudes in children, that is to say on the basis of structures and curricula which do not prejudge what society wants in a manner favourable to existing élites."

> (OECD, 1975, p. 8)

This was a time, Papadopoulos (1994, p. 73) believed, when the 'social' came to dominate the 'economic', within member countries.

The latter era was marked by the first Ministerial conference on education in 1978, Future Educational Policies in the *Changing Social and Economic Context* (OECD, 1979), held when youth unemployment was becoming a major issue everywhere. The apparent breakdown in the nexus between education and work strengthened the economic imperative in education, and placed education firmly within the discourse of human resource development for economic growth. Nevertheless, the press for an education agenda which was not constricted by economic imperatives, continued. The Secretary-General in his opening address to the Conference remarked not without ambivalence "while it now seems that the role of education in reducing economic and social inequalities may have been exaggerated, this does not mean that efforts to improve the educational opportunities of the disadvantaged should be abandoned" (OECD, 1979, p. 11). The Education Committee, too, was keen to dampen down unrealistic expectations about the role education could play in solving youth unemployment. In 1982, it issued a formal statement declaring, *inter alia*, that:

> " . . . expectations should not be raised about the ability of education and training to resolve problems which are primarily the responsibility of other fields of policy, notably the creation of jobs through economic and employment policies."

> (cited in Papadopoulos, 1994, p. 150)

Still, the squeeze was acknowledged. In 1982, CERI's mandate was amended to "emphasize a specific OECD approach to educational problems in relating them to the economic and social objectives of Member countries" (OECD, 1983a, p. 60). The intention was to "establish closer links between the educational activities and other work within the Organisation" (*ibid.*). The Secretary-General's report for that year stated:

> "The central issue to which these programmes [the Education Committee and CERI] were addressed is how, in a situation of cut-backs in public resources, the dynamism

of the educational sector can be maintained so as to satisfy new social demands brought on by deep structural changes in the economy and society. Member countries are increasingly dependent on highly-skilled and adaptable populations capable of promoting technological change, industrial development and social progress. Education and training have to respond to this challenge at a time when young people face the prospect of unemployment."

(OECD, 1983a, p. 61)

A year later, the Report developed a similar theme though differing in emphasis:

"It is also clear that, over the longer term, education cannot be planned solely in response to economic needs and that it has to take into account the new social and cultural developments which are taking place in parallel to the changing conditions of the economy."

(OECD, 1984, p. 51)

Education and the Economy in a Changing Society, the topic of an intergovernmental conference held in 1989, a decade after the first Ministerial conference, set a more explicitly instrumental framing for the mix of social and economic concerns. While the Secretary-General insisted that the conference would discuss "much more than conditions of economic efficiency" (OECD, 1989b, p. 8), the opening address by the conference chair, Australia's then Minister for Employment, Education and Training John Dawkins, reflected prevailing human capital assumptions of links between investment in education and economic growth:

"We acknowledge that education is but one of many factors which contribute to the performance of our economies and labour markets ...   We accept, pragmatically, that the relationship between economic performance and human capital investment can never be measured with any precision ...   We recognize also, however, that the world's most successful economies over the past two decades have given a high priority to education, skills and training as vital factors in their economic success, and have framed their policies accordingly."

(OECD, 1989b, p. 11)

As Papadopoulos observed (1994, p. 171), the renewed interest in human resource development was not new, but notable "for the force and urgency with which educational change was *politically* advocated to respond to the new economic imperative, marked by growing country interdependence and competition in the global economy". Such considerations gave added impetus to 'horizontal' analyses of structural change within the OECD. They encouraged a more comprehensive treatment of issues, the outcome of which was to integrate educational work even more firmly into the mainstream of OECD thinking. At the same time, such integration laid bare the inherent tensions between social and economic purposes of education. The OECD's statement on a "New Framework for Labour Market Policies" (OECD, 1990b) paid particular attention to the role of education in helping to resolve employment problems — a direction which, once again, caused the Education Committee to issue its own *Statement on Education and Structural Change* which began:

> "The interaction between education and social and economic restructuring is a central concern of the educational activities of the Organisation. The importance of education, training and human resource development in facilitating such restructuring is clear. So, too, is the fact that this interaction is not a simple process of merely adjusting education to meet changing social and economic needs. It is rather one which underscores the role of education itself, through its direct impact on individuals ... and the multiple objectives which it serves."
>
> (OECD, 1989c, p. i).

For all that, the notion of structural change clearly signalled a changing discourse in educational reform, one identified with a more efficient state in a global economy and opening up a new way of thinking about the linkages between education, the state and the economy. Apple (1992, p. 127) argued that the significance of the 1989 Conference lay less in the details of the discussion as in " ... the overall orientation of its analysis and its linguistic strategies in creating a rhetoric of justification for a tighter connection between educational systems and the world economy". While attention was paid to more general purposes of education, they were, Apple remarked, "almost always seen against the backdrop of a crisis in productivity and competition."

By the early 1990s, the Education Committee supported a view of education associated with the neo-liberal interpretation of the state which was vigorously promoted at that time within the 'parent' Organisation:

> "It is now becoming evident that the model of publicly-provided, publicly-controlled schooling for all has limitations which can be overcome only by taking a wider, more flexible and more creative view of the educational process itself and its place in our social lives ... We may be coming to the end of the era in which the concept of the 'public interest' in education is best served by the elaborate and costly machinery of ever multiplying, more or less centralised public structures and institutions. The reason is not only cost. More fundamentally, it has to do with evolution of ideas about the proper sphere of action: for government at different levels, communities, enterprises, professional bodies and individuals."
>
> (OECD, 1992d, pp. 7–8)

The shift in orientation anticipated a very different setting for education's social and economic purposes, which was expressed not only in giving greater weight to links between education and the world of work, but also in the privileging of more 'distant' modes of educational governance. From this came new activities for the Education Committee and CERI, on accountability and governance, vocational education and training, lifelong learning, and mass tertiary education. In short, during the 1990s a sea change took place in the way social and economic concerns in education were articulated, a change simultaneously subtle and obvious — obvious because of the new language of business and trade; subtle on account of the continuities in projects and activities. Thus, the 'changing role of vocational education', slowly transmuted into 'transition from school to work' with different emphasis evolving over time. CERI work on teacher quality continued but changed into a study on 'teachers and students: innovative combinations of learning and work', cast in the new language of enterprise and production which looked at "students as producers" and as "workers" and to consider the implications of such "work activities" for, amongst other things, the concept of the "learning enterprise" . . . (OECD, 1993b, p. 15).

Continuities and discontinuities will be developed in more depth in relation to: equity and educational disadvantage; and quality. These two topics have been chosen because they reveal different dimensions within changing priorities as the pressures of globalisation begin to bear down during the 1990s.

Shifting Discourses of Equity

Before discussing changing OECD approaches to educational equality, it is appropriate to say something about the idea of equality itself. Increasingly it is held that the idea of equity does not have a single essential meaning. Rather it represents discourses that are historically constituted and able to be used to promote conflicting and divergent political undertakings. The ideal of equity assumed different forms at different times. Liberal democratic views, drawn from writings of philosophers such as John Rawls (1972), emphasised a concept of equity as fairness, meaning the most extensive basic individual liberty compatible with similar liberty for others. Such a view promoted equal opportunities-oriented programs and policies intended to remove barriers which arose out of unequal power relations which in turn prevented access and participation. Social democratic views of equity, on the other hand, emphasised 'needs' based on a collectivist and cooperative image of society and policy focused on institutional reform rather than on individual access and on improving outcomes for disadvantaged groups.

More recently, in most OECD countries, and certainly in the course of the 1990s, the idea of equity has been restated in line with the requirements of a market-driven economy and more in keeping with competitive individualistic ideology. With the restructuring of the state, equity considerations have become secondary to imperatives of reshaping institutions to make them responsive to market considerations, more efficient and effective and able to compete effectively within a global economy. Under this market-individualist view equity is still important, but is detached from the issues of social redistribution and linked to the issues of people's entitlements to what they produce. A market logic is introduced, converting citizens into consumers, their relations with the state are now mediated by concerns of their productivity and their contribution to national economic growth. In policy terms, this rearticulation represents a cultural change which implies a marketisation of education in which consumers' right to choose becomes paramount. Applied to the public sector, these market principles highlight the importance of cost-efficiency of services, measurement of performance and, wherever possible, commercialisation of services. Equity is set within a discourse of individual rights and capacities, rewards, deserts and entitlements.

From early in its history, the OECD displayed a consistent interest in questions of equal educational access, participation and outcomes, expressed in different policies where the varying objectives of equality, considered above, competed with each other. The range of its interests and the shifts in the discourses of equity which underpinned its work can be gleaned from published titles, for example: *Education, Inequality and Life Chances* (OECD, 1975); *Integration of the Handicapped in Secondary Schools: Five Case Studies* (OECD/CERI, 1985); *Girls and Women in Education* (OECD, 1986a); *Multicultural Education* (OECD/CERI, 1987); *Disabled Youth: The Right to Adult Status* (OECD/CERI, 1988); *One School, Many Cultures* (OECD/CERI, 1989); *Disabled Youth: From School to Work* (OECD/CERI, 1991); *Women and Structural Change* (OECD, 1994c); *Education and Equity in OECD Countries* (OECD, 1997b). While much of this work was set in terms of employment and employability, thus reflecting the way education was legitimated within the Organisation, a wider terrain was explored up to the 1990s, which included for example the rights of disabled people, or good practices in multicultural education. These broader excursions reflected the mix of liberal democratic and social democratic views of equity and educational purpose reflective of the OECD membership generally.

Over the past decade or so, however, that conceptualisation of equity has changed, becoming increasingly couched in terms of the dominant discourse of human resource development and the new individualism. New interpretations of human capital theory are micro- rather than macro-oriented. Rather than seeing education as the 'residual factor' in economic development, human capital theory focussed on the pay-off to both individuals and societies of investment in the formation of particular skills for specific labour market niches. As Marginson (1997a, p. 155) noted in respect of the OECD's report, *Education in Modern Society* (1985c), what becomes emphasised is the need for "flexible, adaptable and self-managing workers" able "to recognise opportunities to change their jobs or to seek new ones, develop their skills and improve their incomes" (p. 155). Such an interpretation supports both liberal democratic and market liberal readings of equality — it promotes, respectively, a policy concern with equality of opportunity (for individuals), or with equity as an individual right. The discursive reframing of equity, in terms of human resource development — that is, as a means to utilise more effectively the available pool of human talent — allowed the notions of social obligation and market efficiency once conceptually distinct to be linked together. Within the Organisation — and elsewhere — the harnessing of equity to efficiency was seen as unproblematic though, more broadly, critics have argued that each pulls in different directions (Henry and Taylor, 1993).

After 1993, explicit equity-related activities within CERI and the Education Committee lapsed (with the exception of its work on disabled people) as equity issues became enfolded into mainstream projects which concentrated on mass tertiary education, vocational education and training, and lifelong learning. *High-Quality Education and Training For All* stated matters thus: "To make equity concerns a foremost policy priority can be viewed as a realignment of aims after a period when qualitative, efficiency and economic criteria have been especially prominent" (OECD, 1992e, p. 89). The rationale of equity was clearly set forth:

> " ... The benefits of learning should be extended *to all* — the long standing equity concerns combine with the compelling economic fact that OECD countries simply cannot afford to allow large pools of their potential talent to lie unexploited. Together these argue for the position and interests of the educationally under-represented to become prominent policy targets in the 1990s. ... "
>
> (OECD, 1992e, p. 99)

The coalition of efficiency and equity goals was reiterated in *Education and Equity in OECD Countries* (OECD, 1997b):

> " ... expectations are again high that through learning, knowledge and retraining, an imposing set of problems in the economy as well as society in general will be alleviated. Applying such learning to the labour market, far from representing a withdrawal from equity in favour of efficiency, is an integral part of its realisation — equity and efficiency can co-exist, indeed must."
>
> (OECD, 1997b, p. 8)

Tying together equity and efficiency in this manner had ambiguous effects. On the one hand, harnessing equity to the mainstream agendas reaffirmed the secretariat's view of the importance of educational equality. Thus, countries choosing to draw on OECD analyses of equity issues would have some rich pickings. On the other hand, the commitment to equity could be read as largely symbolic given the broader discursive context in which educational work was then embroiled: namely, a post-Keynesian policy consensus which anticipated a lesser role for the state, more distant forms of state steering expressed in various forms of educational devolution, of increasing use of 'user-pays' funding mechanisms and the creation of more competitive regimes in education — in short, a climate of increasing deregulation. Here lay the source of tension, for a good deal of research suggests that deregulatory regimes do not deliver equity outcomes particularly well. For instance, it has been shown that educational devolution in the context of a competitive market increases rather than decreases inequalities within and between schools (Gewirtz et al., 1995; Whitty et al., 1998; Blackmore, 1999; Lauder et al., 1999), and moreover the formation of competitive training markets seems to reduce rather than increase training opportunities and outcomes for under-represented groups (Pocock, 1992; Barnett, 1993).

This tension was certainly recognised within the OECD. The report *School: A Matter of Choice* (OECD, 1994d), pointed to problems with choice policies working within competitive systems and stressed the responsibility of educational authorities to provide "an even distribution of educational opportunities" (p. 50) within more deregulated contexts. Drawing on the experiences of a number of countries attempting to do just this, some strategies were noted. *Women and Structural Change* (OECD, 1994c) called for a stronger role for the OECD in monitoring how far equity objectives had been achieved in member countries:

> " ... The OECD has an existing mandate to conduct multilateral surveillance of structural reform. The monitoring process — including gathering and disseminating information and conducting regular country reviews — should apply to the performance of member countries in meeting equality objectives in the context of structural reform. Including gender equity within the agenda of structural adjustment requires that those issues also penetrate the work of the various OECD committees dealing with this matter."

> (p. 44)

However, the proposal to include equity performance indicators in country reviews was not taken up, nor was the proposal to incorporate equity expertise directly into the various OECD committees dealing with structural reform. Indeed, it was resisted (IN 5: secretariat, 1995). Consideration of the management and implementation of equity — as distinct from analyses of equity issues — appears to be one agenda item led by active member countries rather than the secretariat (IN 5: secretariat, 1995). Against this background, it is perhaps not surprising that a later publication, *Education and Equity in OECD Countries* (OECD, 1997b) was silent on matters of the regulatory framework in which equity objectives set out in that report might be achieved. It was equally silent on the potentially deleterious effects of system devolution — a policy strongly supported by the Organisation — on equity outcomes.

One explanation for this analytical blind spot may be that as equity was drawn into the main economic arena the OECD became, as it were, discursively stuck. With the deregulatory genie out of the bottle and in the context of a more explicit economistic framing of educational agendas, the critical discursive capacity to interrogate the mantra of 'efficiency with equity' vanished — even though outside the OECD, and, in principle, accessible to it, was a considerable body of feminist and other critical research that pointed to the marginalisation of equity concerns when attempts were made to conjoin equity and efficiency. Nevertheless, following the commitment to equity set out at the Ministerial conference on 'High Quality Education for All', the development of equity indicators was listed as a priority for the 1997–2001 phase of the educational indicators project (OECD, 1996e). The persistent lobbying of some member countries (IT 17: secretariat, 1997) may have paid off.

However, the issue here is not simply a matter of monitoring or regulation. There are also questions about what form regulation and monitoring may take given the new modes of educational governance. In exploring some of the tensions at play we turn to another cognate issue, quality.

The Quality Agenda: Entangled Discourses of Improvement and Accountability

While the notion of quality is at one level uncontentious, even banal, it has emerged as "one of the most important policy issues on the institutional and political agendas in Europe" (Maasen, 1997, p. 111) and elsewhere. Especially in higher education, it assumed the status of a global discourse, having impact not only on the governance of national systems, but also consolidating an international higher education market. Quality is redolent with competing meanings which play out in different ways in different contexts, a point noted in an early OECD document this subject:

> " . . . for some it appears to serve as a synonym for excellence or efficiency, others use it as a metaphor for good educational practice and others again equate it with material provision. For many it is no more than a short hand way of expressing value discontent with the present outcomes of education while covering up a lack of cogent policies and priorities for action . . . Quality will always remain a subjective entity."

> (OECD/CERI, 1983, p. 3)

Sachs (1994) identified two competing models of quality: quality improvement and quality assurance. The former, she suggests, is a negotiated, consensual model with an internal locus of control, operating through processes of peer review and utilising qualitative indicators of success. By contrast, quality assurance operates within a competitive context. It is driven by government directives utilising processes of external audit based on quantitative indicators of performance. Quality improvement, she suggests, focusses on improvement, while the focus of quality assurance is accountability. Both models can be seen in the OECD's work on quality with the latter becoming dominant in the course of the past decade, particularly within higher education.

However, let us begin with schooling given that, within the OECD, it was at this level where quality concerns first surfaced. Quality emerged as a significant theme for the OECD in the 1980s, heavily influenced by the US Department of Education and following the 1983 report *A Nation at Risk* which compared US student performance unfavourably with that of Japan. The US 'pressed powerfully and persistently' for quality to become a top priority in the OECD's work (Papadopoulos, 1994, p. 181). The organisation's early work on this theme was linked to the effective schooling agenda as part of the move to system devolution and towards various forms of school-based management:

> "The effectiveness of schooling, particularly at compulsory school level, has become a central issue in the political debate about education in all Member countries. A new activity Quality in Basic Education resulted in an initial series of national reports on

how quality is defined in different countries, the main factors which determine quality and how educational outcomes are assessed."

(OECD, 1985b, p. 51)

Quality framed in this context was inextricably tied into a whole range of contentious issues which surrounded questions of curriculum and pedagogy, teaching and learning, parental choice and the development of accountability mechanisms in decentralised systems. Debates about quality were in fact ideologically charged and revolved around (at least) two axes. One pertained to educational purposes. The other related to the means of achieving them.

Educational purposes: tensions between quality and equality

Debates around the meaning of quality, in Papadopoulos' account, reflected the ideological cleavage within the Organisation between egalitarians (largely following a Scandinavian model) and Thatcherite/Reaganite neo-conservatives:

"Reduced to its essentials, the debate was about the fundamental purposes of schooling: the one camp seeing them as essentially concerned with learning outcomes, assessed in terms of subject matter mastery, particularly in the traditional basics; the other viewing such outcomes within a broader social educative role for the school ... "

(Papadopoulos, 1994, p. 182)

In procedural terms, he suggested, this rift led, on the one side, to an emphasis on choice, competition and privatisation — i.e. the application of market principles — and on the other side, to an emphasis on more child-centred approaches and pleasant learning environments (p. 182). The contrasting viewpoints and tensions surrounding the issue of quality were reflected in the submissions to the 1984 conference 'Quality in Basic Education'. The position taken by the UK, the US and Denmark was contested by the other Scandinavian lands — resulting in "lively debate" and "controversy" which was "unusual for this type of meeting" (p. 183).

Still, the Organisation went on to identify a number of problems around which member countries could work in a more pragmatic manner — for example, teaching methods, the curriculum, school organisation, assessment and evaluation. It also drew up a number of common concerns. Thus, the first substantive publication on the theme of quality, *Schools and Quality*, noted the changing terms of the educational debate:

"The traditional contrasting ideologies of educational debate — conservative versus progressive, elitist versus egalitarian, right versus left — are increasingly inadequate terms to apply to today's complex debate and the concern for improving quality spans these different divides."

(OECD, 1989d, p. 136)

Common concerns were identified as: a reassertion of qualitative considerations arising from the expansion and widening access of education systems; a concern to improve the quality of schooling for disadvantaged students; and the importance of human capital in economic development — in training and recurrent education as well as in schooling.

Schools and Quality informed the third Ministerial Meeting on Education in 1990 on the theme 'High Quality Education and Training for All''. There, notions of quality and equality were brought together:

"The emphasis on quality underlines the key role that learning plays in sustaining economic, social, cultural, and political well being. That learning of quality should be 'for all' is to recognise that education is the corner stone of participation in our societies."

(OECD, 1992e, p. 3)

The report from the meeting pointed to the complicated mix of factors that underpinned the interest in quality: a reaction against "excessive egalitarianism" and the "sacrificing" of standards (p. 56); a more conservative ideological climate in favour of a greater role for market forces in education, the concomitant rhetoric of choice, accountability and competition; a more generalised tendency to blame education for a sense that 'all was not well'; the ever-expanding brief for education against which traditional institutions tended to appear 'monolithic and conservative'; the declining status of teachers. In short:

"This combination — growing expectations of education alongside unprecedented criticisms and the declining recognition of the professionals charged with carrying out these responsibilities — suggests that the concern for raising quality is, in important measure, an expression of malaise and confusion about educational ends and means."

(OECD, 1992e, p. 57)

Nevertheless, the report took a highly optimistic view of the possibilities for education, and pointed out parallels between the 1990s and the 1960s. It concluded that the process of:

" . . . identifying the ends and means for offering high quality education and training for all is to emphasize the equity concerns so prominent in that earlier watershed decade. But instead of 'quality' and 'equality' being sharply contrasted as commonly occurred then, the aim should now be to bring these two broad aims into full harmony."

(OECD, 1992e, p. 63)

Some understanding of what was meant by "full harmony" emerged in the treatment of the educationally "disadvantaged and under-represented" whose exclusion, it was argued, entailed economic, social and individual costs. Having noted the plethora of policy responses for dealing with educational disadvantage, the document concluded that:

" ... the most effective strategies will have as their main focus the mainstream provision of education and training where the aim is to provide high-quality learning to all ... The aims of quality and equality come together ... Both the broad aims of raising quality and of increasing equality point to the need to develop a culture of learning ... An intensive high-quality start, with well-resourced schools imbued with a culture of learning, can reap larger dividends over the long term than a wider extension of mediocre provision."

(OECD, 1992e, p. 96)

The problem with such a formulation is that the discursive linking of quality and equality operates at so high a level of abstraction as to be virtually meaningless. For example, the rhetorical flourish towards "a culture of learning" as the means to link quality and equality and invoking well-resourced schools ignored the sociological evidence — which the OECD had itself compiled over the years — of the deeply embedded social and cultural divide in education that worked against the easy achievement of such a goal. The foreword to the report acknowledged contextual factors such as increasing social division which flowed from economic globalisation and reduced government support for education. The quality rhetoric, however, underplayed the real difficulties likely to be encountered in such circumstances to develop "cultures of learning" or to provide "well-resourced schools".

What were, then, the mechanisms for achieving quality outcomes? Here, the notions of quality as improvement and quality as accountability — or quality assurance — allow a useful way to examine this issue.

Tensions between quality improvement and quality assurance

In essence quality *improvement* entails more qualitative assessment of institutional performance and is associated with a broader social agenda. Quality *assurance* relies more on quantitative measures of performance and underscores economic imperatives. Something of this distinction came through in the study *Schools and Quality* (OECD, 1989d):

"Two potentially conflicting ideas tend to be confounded in the current interest in quality: on the one hand, that there is a shift in priority from inputs (including financial resources) to outputs; on the other that the complex chemistry of schooling defies the application of simplistic input–output models. It would be mistaken to argue that a revived focus on outcomes implies that the inputs and processes that determine them become accordingly less important; both are crucial."

(OECD, 1989d, p. 135)

The OECD's work on quality assessment over subsequent years displays a mix of improvement and of accountability. While they pulled in different directions, the OECD argued that the approaches were not necessarily dichotomous in view of the links between school effectiveness and new modes of devolution, for example, which carried with them greater demands for external accountability.

The potential linkages between school improvement and accountability processes were brought out in *Schools Under Scrutiny* (CERI, 1995a), the report of a study in the CERI series 'What works in Schools?'. It looked at approaches to evaluation of school performance in seven OECD countries. In outlining the background to the study, the report noted:

> "Many OECD countries, strongly influenced by economic doubts and difficulties, have been reassessing the quality of their school systems, looking at how far they succeed in educating the young to the maximum extent in the light of increased economic competition with other countries."
>
> (CERI, 1995a, p. 13)

Thus, as the report made plain, lying behind the general concern with "quality and relevance" were a number of factors, including "a stronger voice for the users of the education system, more choice and competition, a devolution of responsibility to schools, and a new emphasis on accountability" (p. 14). The 'key rationales' for school evaluation, it concluded, are accountability and school improvement (p. 21). A good deal of discussion in this report linked accountability issues to school improvement. At the same time, the report cautioned that accountability "as a concept does not necessarily address the improvement process itself" (p. 24), and that school evaluation was not a 'magic wand' for school improvement (p. 46).

The study noted a variety of evaluation approaches. Some leant towards external accountability mechanisms: assessment of individual teachers in Germany; school inspections in England and New Zealand; the use of performance indicators, mainly student achievement results, in the US. Others tended towards more qualitative assessments of school improvement, such as the Swedish focus on "friendly inspection and self improvement" (p. 44). The report itself was critical of some of these developments. It opted for qualitative approaches to school improvement grounded in principles of reflection and self review: "Not only is this type of self-review likely to be more cost effective (i.e. both cheaper and more effective) than some of the more elaborate accountability mechanisms, but through it schools can truly become learning organisations" (CERI, 1995a, p. 47).

Self-reflection also permeated CERI's work on teacher quality, culminating in a study entitled *Quality in Teaching* (CERI, 1994a). In this report, quality was again conceptualised broadly, with a strong focus on school improvement. Qualities of teacher commitment, empathy, collaboration and self-reflection were identified as key aspects in quality teaching, along with managerial competence, pedagogic skills and curriculum knowledge. Although some studies in the report focussed on performance appraisal, for example the US contribution, this approach was not given much emphasis in the report.

This qualitative approach to school reform was summarised in Townshend's review of OECD work on the quality of teachers and teaching:

"The message for policy makers is that policies should support the growth of local, school level, developments rather than attempting to mandate change in detailed central regulations. The aim should be to produce policy active schools which feel responsible for the innovations which are necessary in order to carry out their own policies, to meet the particular challenges they face in their social context."

(Townshend, 1996, unpaged)

Previously, work on quality assessment involved a shift towards examining the performance of educational institutions, but set largely in terms of quality improvement's emphasising qualitative assessments within local contexts. It was an approach that contrasted markedly with that taken up in another report *Performance Standards in Education: In Search of Quality* (OECD, 1995b) published around the same time. In the latter report, quality was interpreted narrowly as 'standards'. This study, conducted in 1994 by CERI and the Education Committee 'at the request, and with the cooperation and financial support, of the government of the United States" (OECD, 1995b, p. 3) was "facilitated by the secondment of an official from the Office for Standards in Education in the United Kingdom" (p. 3). Its focus was on improving methods of setting, applying and monitoring standards of student performance against the backdrop of the accelerating movement towards school devolution and the concomitant demand for public accountability across education systems in all member countries. In such a context, quality was defined as a system outcome, linked to the broader project of performance indicators in education and to mechanisms of quality assurance. It is this approach that has dominated the Organisation's approach in higher education, particularly through the activities of the Programme on Institutional Management in Higher Education (IMHE).

Quality in higher education and the ascendance of performativity logic

Within IMHE quality became an issue in the early 1990s, and subsumed its earlier work on performance indicators. The establishment of national quality assessment systems across many countries was described by IMHE as "part of the changing relationship between higher education institutions and the state" — a relationship characterised by a mix of accountability and evaluation mechanisms, heavily dependent on peer review and institutional self-evaluation, replacing systems of either close government regulation or high levels of institutional autonomy (OECD, 1994e, p. 6). In the early 1990s, quality assurance became the dominant management ideology for addressing pressures felt across education systems in all OECD countries, in particular, those which accompanied the shift from elite to mass tertiary education. Harnessed to this development were other elements: the interest in school effectiveness (with all its contested meanings) mutated in higher education into concern with institutional improvement and self-evaluation; the wide-ranging preoccupation with accountability as a form of indirect steering expressed through new management theory; more general disquiet transmitted particularly, though not solely, from the US about standards and institutional performance; unease about academic equivalence and institutional comparability before the enhanced mobility of staff and students and an increasingly internationalised labour market (Green, 1994). All were present.

These elements have come together around a so-called "global model of quality", urged on by a goodly number of international agencies and organisations including the OECD, UNESCO and the World Bank (Lenn, 1994). The model applies both internal and external modes of assessment, the main elements of which include: meta-level assessment independent of government; self-evaluation; peer review including site visits from external experts representing appropriate constituencies; non-ranked reporting of results which may be published or confidential; non-linked relationship between performance and funding thus avoiding a "compliance culture" (Van Vught, 1994). A number of national and international quality assurance agencies were created, for example the British Quality Assurance Agency, the New Zealand Qualifications Authority, the Danish Centre for Quality Assurance and Evaluation of Higher Education, the quality assurance project of IMHE within the OECD, and the International Network of Quality Assurance Agencies. In higher education quality assurance provides a good example of how educational governance has moved upwards and downwards since much assessment is carried out either within institutions and/or via such agencies some of which work independently of, or distant from, governments.

While there are many accounts of differing approaches towards assessment of quality (e.g. Craft, 1994; Bauer and Kogan, 1997; Maasen, 1997; Vidovich and Porter, 1997), our interest here is less in the details than what the move towards quality assurance may represent *in toto*. In general, our contention is that widespread interest in quality assurance is symptomatic of a particular mode of governance which in its turn derives from the logic of what Lyotard (1984) called "performativity". Performativity embodies the ascendance of a technicist mentality embedded in a system of governance which brings together the features of both Weberian iron cage and the Foucauldian panopticon. The logic of performativity generates highly refined mechanisms of surveillance and elaborate accountability mechanisms. One consequence, as Neave (1988) pointed out in his discussion of the evaluative state, was to shift the focus of administrative attention from process to product control, i.e. from inputs to outputs. Quality assurance in such a context tended therefore to focus on quantitative measurement rather than on the qualitative evaluation of performance.

Ball (1997, p. 5) qualifies performativity as a mechanism of indirect steering "which replaces intervention and prescription with target setting, accountability and comparison". It poses questions in relation to quality assurance such as what kinds of quality targets may be set and what kinds of performances may be compared. At one level responses to such questions will be highly diverse given differences in local circumstances, institutional cultures, systemic priorities. Indeed, a variety of approaches is implicit in the global model of quality assurance.

Yet, the mere existence of a global model suggests a more convergent dynamic, a manifestation of what Davies and Guppy (1997) term global rationalisation. Global rationalisation and economic globalisation may be conceived of as independent, but nevertheless linked. Taken together these phenomena may help to explain the increasingly common form of the contemporary state, variously described as efficient (Cerny, 1990), corporate (Yeatman, 1990) or managerial (Clarke and Newman, 1997), and its increasingly common set of educational priorities: the cost-effective delivery of educational services; a more competitive educational market driven by principles of consumer choice; and the production of high quality human capital. Linking of such priorities to a particular mode of governance constitutes what Peters and Marshall (1996) call "busnocratic rationality" in education, that is, an:

> " ... emphasis on values which are inherently attached to business and industry but which can be overlaid and injected into higher education through deregulated privatisation processes which merge culture and commerce."
>
> (cited in Meadmore, 1998, p. 36)

Against this backdrop the ideological tensions surrounding the quality agenda are played out around notions of improvement and accountability with the latter now ascendant.

We are not proposing a deterministic relationship between the logic of performativity and quality defined in busnocratic terms. There may well be instances when quality assurance attempts to assess innovative teaching approaches which may be costly and resource intensive, rather than linked to cost effective use of new technologies as means of teaching large classes. In addition, account needs to be taken of different pressures exerted by dimensions of globalisation other than the economic — for example for management of cultural diversity as an offshoot of cultural globalisation. Quality assurance mechanisms may well be deployed to assess institutions' performances in the development of inclusive curriculum, for instance. In a sense, however, the targets of quality assessment are almost irrelevant, though this argument should not be driven too far. Feminist and Indigenous groups, amongst others, have made good use of quality processes to argue for system compliance to equity objectives. But essentially, quality assurance with its linguistic and procedural resonances with the world of business, in its technicist instrumentality and in a context of declining state support for higher education, represents a hollowed-out vision of educational purpose in which image outweighs substance (Blackmore, 1999, p. 154) or, in effect, means become ends. It is this aspect to which Neave (1993) was alluding in his acerbic comment: "the issue behind quality has very little to do with quality ... Quality ... is a technique which allows a national administration to insist on ends while rigorously denying the means" (cited in Dias, 1994, p. 166).

Take the examples of publications outputs, often regarded as a significant indicator of institutional quality (Dill, 1998), or the various forms of student evaluations of institutional performance such as the Australian Course Experience

Questionnaire. While there is much cynicism and bickering about the validity of quantitative publication measures (how many small multi-authored scientific articles equal one 'real' single authored essay, and so forth), or about the conditions under which questionnaires are conducted and interpreted, academics are driven by performance demands, forced by the surveillance mechanisms of the quality assurance panopticon to compile ever-more manicured profiles to demonstrate improved personal and institutional performance and help both to sharpen institutions' competitive edge and to justify their continued (albeit reduced) funding. In the process, selected pieces of evidence of 'high quality' performance are inserted into glossy brochures or web site text to lure the student dollar, win corporate sponsorship, seal collaborative ventures and other revenue-earning tactics.

This is not to suggest that quality assurance in higher education equates directly with goals of effectiveness and efficiency, though institutions may seek to measure it. Rather, quality assurance mechanisms enable three objectives to be realised: tighter control over activities governments regard as really important (the steering may be distant, but very firm); a high degree of manufactured compliance around stipulated quality targets (statistics may be massaged to produce the picture required); and the creation of league tables at national and, increasingly, international levels (with institutions or systems selectively promoting those features in which they have demonstrated superior quality performance). What counts as quality in the current circumstances is more likely to favour an economistic view of education rather than viewing it as a public good or as an intellectual activity. Despite the rhetoric of self-improvement and non-competitive evaluation outlined in the global model, quality assurance has of necessity become part of the competitive apparatus of institutions, in turn linked to the imperatives of image management (Blackmore, 1999) which accompany the process of global commodification of any product, higher education included.

Little critical edge is to be found in the OECD's treatment of quality despite acknowledging problems in conceptualising quality and in implementing quality assurance processes. Take, for example, the treatment of quality in the publication which followed upon the thematic review of the transition from elite to mass tertiary education, *Redefining Tertiary Education* (OECD, 1998c). Difficulties encountered with the quality-as-accountability approach were highlighted:

> "To date, much of the interest in quality appraisal and improved management arises from requirements for transparency and accountability and from the increasingly entrepreneurial roles of institutions. Equally important are appreciations of the quality of the educational experience of students and of the capacity of institutions and whole systems to provide conditions for a high standard of teaching and learning for all students, that is, quality appraised for the purpose of educational development."

> (OECD, 1998c, p. 69)

From this starting point, a number of learning-based problems were identified: attrition rates, lack of interest amongst tertiary teachers in student learning particularly in mass higher education, the high cost of remedial measures, the tendency by institutions to over-enrol and thus to overcrowd, the general issue of acceptable standards. National propensities were noted, for example, the UK preferred scaled evaluations and league tables whereas other countries opted for institutional self-improvement. However, since any analysis of why the entrepreneurial model of accountability loomed so prominently on the quality agenda was absent, the assertion of the need for alternative questions to be directed to matters of educational improvement seems weak. For example, the importance of strengthening the place of teaching development and support units in universities was recognised, but little acknowledgment was paid to the current logic which ties such units into an agenda of technologised teaching for large classes, in effect promoting a type of pedagogy which may well aggravate rather than solve, problems of attrition. Problems of "burdensome documentation" and increasing administrative costs (p. 74) were touched upon, but without questioning the underlying logic of performativity — and still less the associated danger of converting means into ends — which drives the quality assurance agenda. The absence of a critique may spring from confounding quality concerns with institutional management — and by extension with managers perhaps less sensitive to the slippages between assessment of quality and the day to day life of institutions whose performance is ostensibly being measured. This lack of critique reflects, we would argue, the extent to which the Organisation is discursively blinkered.

As has been indicated earlier in this chapter, a new perspective for education is gaining momentum around the issues of social inclusion and exclusion. Whilst this topic will be pursued more extensively in the final chapter, it is worthwhile to speculate about where such framing might lead.

A New Framing for Education: Social Inclusion and Exclusion

This chapter has shown how over time social and economic concerns have played off against each and how they have been rearticulated within the OECD's educational work. Most of this work stresses the complementarity of social and economic goals but it is placed in a context of persisting (or worsening) structural inequalities within OECD countries which drives against the provision of high quality, equitable education for all. In such a context, scarcely any critical examination has been made of the rhetorical link between equality and efficiency, or between equality and quality. However, the disquiet voiced in the quotes at the head of this chapter reflects a growing consciousness about the threats to political and social cohesion to which neo-liberalism gives rise. It is possible, therefore, that the interest in social inclusion/exclusion, which surfaced during the 1990s as the socially destructive and destabilising outcomes of economic globalisation became evident, could herald a new framing, a new way of articulating the social and economic purposes of education. As one secretariat member remarked:

"Education now has to adapt itself to addressing the problem of social exclusion — a primary role in supporting democracy. At the same time, the economic sector has to address issues of developing the human potential — and not just at school level."

(IN 7: secretariat, 1995)

Interest in social inclusion and exclusion is highly Eurocentric, reflecting the social democratic leanings of the European Union and Prime Minister Blair's "third way" politics of rethinking social democracy in the context of globalisation (Giddens, 1998). The terminology, adopted by both UNESCO, and the OECD as well as the EU, may possibly represent a new way of describing educational purposes. Thus a new language emerges, that of participation in civil society, and of the consequences of being excluded from that society. Some of this comes through in the concluding pages of *Education and Equity in OECD Countries*:

"Education should not be considered exclusively in terms of meeting other ends (better employment prospects, income chances, greater security), but as promoting social and cultural participation, and thereby contributing to the reduction of marginality and exclusion."

(OECD, 1997b, p. 103)

The year following another OECD publication, *Civil Society and International Development*, opened thus:

"Interest in the concept of civil society is undergoing a remarkable renaissance. In the field of political theory, this concept is currently seen as a potential tool to overcome some of the main theoretical and political stalemates ...   Civil society is central to discussions of democratisation, the rule of law, human rights. While familiar in substance to aid operators for a long time, the notion of 'civil society' has acquired a new dimension in the context of governance and democratisation."

(OECD, 1998d, p. 11)

In itself, of course, such rhetoric may mean as little (or as much) as the highfaluting rhetoric of quality and equality. The central question is how far it connects to broader agendas within the Organisation and in the broader policy community. As a parenthesis, it is worth noting that this document was the fruit of collaboration between the Development Centre of the OECD with the North–South Centre of the Council of Europe and hence out of the neo-liberal economic mainstream. Yet, concerns such as these with their mixed motives, are being channelled into the economic and social mainstream, a development symbolised by including 'respect for human rights' as a key criterion for admittance to the OECD club. The recognition now of the socially divisive consequences which economic globalisation in particular brings about seems to drive a new rhetoric of both social as well as economic investment. Two documents gave tongue to these preoccupations, both published in 1997. The first, *Towards a New Global Age: Challenges and Opportunities* (OECD, 1997c) warned: "the productive turmoil of relentless competitive markets ... are [showing] signs of growing strains on the fabric of OECD societies, in the form of stubbornly high levels of unemployment, widening income disparities, persistent poverty, and social exclusion" (cited in Spring, 1998, p. 163). The second, *Societal Cohesion and the Globalising Economy: What Does the Future Hold* (OECD, 1997d), spelled out more explicitly the Organisation's framing for the ceaseless juggling of economic and social priorities. It noted once again the 'political disenchantment' of income polarisation, continuing high levels of unemployment and widespread social exclusion. The document contended that "By shifting from a social expenditure to a social investment perspective it is expected that considerable progress can be made in transforming the welfare state" (cited in Spring, 1998, p. 167). As Spring commented, "Therefore, besides economic returns, social investment is expected to counter public resistance to a market economy and technological change" (p. 167).

At present then, perhaps all that can be said is that the agenda of social cohesion is highly ambiguous. On the one hand, it stands as a reaction to, and is hence residualised by, the main thrust of global economic restructuring. On the other hand, its rhetoric both signals and reflects a powerful counter-dynamic — a species of 'bottom-up globalisation', connected to far broader social agendas and to new discourses of social capital. How this agenda of ambiguity will fare in the educational work of the OECD is not altogether clear.

5

The Politics of Educational Indicators

"The monitoring of progress and experimentation in systems of education depends heavily on indicators that enable government authorities and other interested groups to judge the context and functioning of education and the results achieved. Education indicators can reveal some of the most critical weaknesses of education systems, and can aid the design of corrective policy."

(CERI, 1993, p. 10)

"International comparisons of educational conditions and performance are now perceived as a means of adding depth and perspective to the analysis of national situations. References to other nations' policies and results are beginning to be routinely used in discussions of education, and comparability now belongs with accountability to that changing set of driving words which shape the current management paradigm of education."

(Alexander, 1994, p. 17)

Over the past decade, the project on international education indicators has become a highly significant part of the OECD's work in education. The annual publication *Education at a Glance: OECD Indicators* is disseminated widely across the OECD countries and elsewhere. In addition, since 1996 it has been accompanied by an analytical supplement which comments in greater detail on selected themes of key importance to governments, policy makers and the public. *Education at a Glance* was first published in 1992, to provide insight into the comparative functioning of education systems in member countries. Thirty-six key indicators gave information in three areas of interest: the demographic, economic and social context of education; costs, resources and school processes; and outcomes of education. Subsequent volumes presented data that reflected both on the resources invested in education as well as on its returns, illuminating "the relative qualities of education systems" (CERI, 1996a, p. 9). By 1998, the original categories had been reorganised and expanded to cover six themes: the demographic, economic and social context of education; financial and human resources invested in education; access to education, participation and progression; the transition from school to work; the learning environment and the organisation of schools; and student achievement and the social and labour-market outcomes of education. The significance of these themes will be tackled later in the chapter.

The OECD argues that this exercise in international comparison is designed to assist the processes of policy formation in member countries and to contribute to the public accountability of education systems:

> "At a time when education is receiving increased priority but, like other areas of public spending, is facing the prospect of limited public funds, understanding better the internal processes that determine the relationship between educational expenditures and educational outcomes is particularly important."

> (CERI, 1995b, p. 7)

This chapter will argue that this view of the purposes and significance of the OECD's work on indicators is understated; this undertaking not only provides relevant and comparative information to member countries but also helps shape their policy agendas and priorities. The focus of our interest is less upon the details of the indicators than in their overall significance. Our contention is that a broader interpretation of the politics of change is associated with the Indicators project, which is based on a particular view about the policy directions and approaches required to reform education. In this way, the project fulfils a normative and legitimation role in promoting what amounts to a global ideology of educational management and change tied into broader public sector reform across OECD member countries.

Thus the Indicators project is a handy illustration of the general thesis of this book: namely the transformation of the OECD's role as policy instrument and forum — serving as a catalyst to facilitate policy development in member countries, assisting processes of policy dissemination, adaptation and borrowing — to assuming the status of an international mediator of knowledge and global policy actor. These 'faces' of the OECD are not mutually exclusive because indicators work at two levels. At one level, indicators assist member countries to clarify and compare their own policy positions; simultaneously, international indicators draw countries into a single comparative field which pivots around certain normative assumptions about provision and performance.

A brief historical sketch of the development of international educational indicators within the OECD will serve to introduce a more analytical discussion of their purposes and effects. Our focus is upon the demands of comparability which potentially are driving countries in more 'corrective policy' directions, a tendency implicit in the quotations at the beginning of this chapter.

A Brief History [3]

The OECD has a long history of compiling statistical information on education. Throughout the 1960s interest grew in the use of statistics for educational planning. The Committee for Scientific and Technical Personnel, for instance, collected cross-national data to identify long-term labour market needs, especially in the areas of science and technology. The data were intended to improve public knowledge and appreciation of both education and science and to support the task of reforming higher education and training. Information supplied by the OECD, it was thought, would assist policy makers to make a political case for greater allocation of funds to programs of educational reform.

Country Reviews served a similar political function. Independent experts sent to discuss educational matters with government officials and other interested educational bodies fulfilled an important advocacy role. They reviewed a country's educational priorities. They commented on educational performance, often in terms that were comparative. The 'confrontation' meetings challenged member countries to increase investment in education as a means of maintaining economic growth, an approach which reflected the older 'macro' human capital dimension in education policy.

Such advocacy, however, was not based on any systematic statistical comparison of performance. OECD statistics recorded 'input' measures. Little effort was made to assess the effectiveness of educational systems, though calls were made for the Organisation to collect such data, especially from the US. One key consideration was to develop reliable strategies for forecasting size and patterns of future enrolments at different levels of education, as well as a model to measure efficiency and effectiveness in the use of resources in all areas of education (Papadopoulos, 1994, p. 50).

In 1964, a conference of European Ministers of Education held in London recommended that the OECD:

> " ... whose work in this field is greatly appreciated, be invited to formulate clearly in a model handbook the various factors involved in effective educational investment planning, so that countries represented *may have a basis for the compilation of comparable statistics.*"

(quoted in Papadopoulos, 1994, p. 50, our emphasis).

This resolution, subsequently endorsed by the OECD Council, provided "the additional political impetus" needed to support the statistical work already being undertaken (*ibid.*).

---

[3] Once again we draw on George Papadopoulos's (1994, pp. 50–54) historical account in this section, and also on *The INES Project (1988–1995) (Historical Brief)*, OECD/CERI (1995a).

Interest in educational statistics and quantitative analysis of resource utilisation and management derived from an enthusiasm in academic circles during the early 1970s for systems analysis, program budgeting, operational research and mathematical model-building particularly in the social sciences. Powerful computers were already at work elsewhere in the OECD developing sophisticated simulation models for economic forecasting. These technologies, it was felt, could also be used for educational policy analysis, and in particular for the calculation of costing particular policy options. Mathematical models could be applied to assess the adequacy of communication patterns within education, and more significantly to various diagnostic functions. With its long tradition in educational testing, the United States was the most enthusiastic of all OECD members about the potential of statistical models for measuring outcomes. Some other countries were also interested.

In 1973, the Working Party of the Education Committee on Educational Statistics and Indicators devised a general framework for educational indicators, intended to measure the impact of education on society. A total of forty-six indicators set out to measure education's contribution to the transmission of knowledge, to equality of opportunity and social mobility, to the needs of the economy, to individual development and to the transmission of values. Whilst the Working Party had little difficulty in specifying these performance objectives, the secretariat itself faced difficulties in working with them in any way that was effective. The objectives were set at a very high level of generality. Clarity about how best to judge their achievement was little indeed. Moreover scepticism abounded about the relevance and value of mathematical models in education, both within the OECD and within the international educational community. As Papadopoulos pointed out, "educational statistics were traditionally input-oriented and no country showed enough interest in gearing up the necessary research effort to relate them to output indicators" (Papadopoulos, 1994, p. 127).

In the course of the 1970s, it became clear that the gap between the promise of the sophisticated models of systems analysis and performance research and their effective application to local and regional priorities in education, was considerable. Some experts raised a range of epistemological difficulties in quantifying educational outputs beyond general pedagogic outcomes, such as test scores. Many expressed ideological objections to the performance modelling of so complex a human endeavour as education which, they argued, should not be reduced to mere numbers. More significantly, the practical problems associated with educational data collection and classification proved more difficult to resolve. No country had the information bases necessary to apply the more abstract mathematical measures of performance. Different countries collected data even on such simple matters as enrolment trends in radically different ways. In 1974, a report produced by a group of experts, *Mathematical Models for the Education Sector*, effectively conceded that some of the technical problems involving the models proposed were insurmountable.

This admission did not halt the OECD's mathematical work in education; it simply slowed it down. A manual for the compilation of comparable educational statistics, known as the 'Green Book', continued to be widely used by those interested in quantitative measures as a useful policy instrument (Papadopoulos, 1994, p. 52). The OECD continued to compile data on input measures supplied by member countries and published volumes on International Educational Indicators in 1974 and 1975, and again in 1981 (Papadopoulos, 1994, p. 190).

Amid continuing ideological and philosophical debates about the nature and applicability of performance indicators to education, the OECD — and CERI in particular — explored issues of educational reform, social equity and innovation in terms more conceptual and philosophical than evaluative and statistical. Within CERI, a culture of distrust towards performance indicators had grown up over the years. By the mid-1980s, however, even CERI could not turn aside pressures for a new initiative to develop indicators. The US, in particular, repeatedly called for work on outcomes indicators, specifically related to school effectiveness, threatening at one stage to withdraw support from CERI if its demands were not met (IN 66: bureaucrat, 1996). Backing came from a different ideological quarter. France — with its bureaucratic interest in statistical data collection — joined with the US in urging the OECD towards the development of educational indicators (IT 3: secretariat, 1996).

In a Presidential address to the Comparative and International Society, Heyneman (1993, p. 375) described a visit made to the OECD in Paris in 1984 in the aftermath of an acrimonious meeting of CERI's board of directors. At that meeting:

> "The US delegate was said to have put a great deal of pressure, and in very direct language, for OECD to engage itself in a project collecting and analyzing statistical education 'inputs and outcomes' — information on curricular standards, costs and sources of finance, learning achievements on common subject matter, employment trends and the like. The reaction among the staff of CERI was one of shock, and deep suspicion. Those whom I interviewed believed it was unprofessional to try and quantify such indicators, and that it would oversimplify and misrepresent OECD systems, and that it would be rejected by the twenty-four member states whose common interests they were charged to serve."

However, the strength of the United States' convictions was such that CERI had no choice but to concede. It bowed to internal pressures within the OECD. It also yielded to the weight of the accountability movements in several member countries, where for some time decision makers had been calling for comparative data to assess and monitor the effectiveness of their education systems. In Papadopoulos's (1994) opinion: "It seemed therefore logical to add an international dimension to these national efforts, even though the difficulties, both conceptual and technical were fully recognised from the outset." Furthermore, since considerable groundwork had already been laid, the OECD was "well placed to respond to the mounting pressure in the late eighties for a new governmental effort to develop such indicators" (1994, p. 190). By the early nineties, according to Heyneman's account, the doubters had been won over and the Indicators project became fully established within the OECD's educational work — a situation which, he argued, reflected a burgeoning "new industry of comparative education" (Heyneman, 1993, p. 378).

Work began cautiously, with an international meeting in Washington in 1987 followed by a second in France in 1988. In May 1988, the CERI Governing Board finally established the International Indicators and Evaluation of Educational Systems (INES) project, which was set up with the initial purpose to do exploratory work on construction of indicators based on existing sources or reasonably accessible new data. Some degree of international validity was also important. The exploratory work was carried out by five networks, each with a leading country, and was guided and monitored by a Scientific Advisory Group, assisted by the secretariat. The five Networks explored indicators on student flows (Australia), student achievements (US), school processes (France), costs and resources (Austria) and attitudes and expectations towards educational systems (The Netherlands).

CERI's position on indicators was reflected in the decision to place the INES project in the charge of a member of the secretariat with a philosophical background in the expectation, perhaps, that it would not succeed (IT 3: secretariat, 1996). However, the exploratory work ended in December 1989 with a surprising degree of consensus and enthusiasm, demonstrating "that it was possible to overcome the initial reluctance" in relation to the project (OECD/CERI, 1995a, p. 1). Against CERI's expectations, "the indicators received a splendid political reception" (IT 3: secretariat, 1996) and the second phase of the INES project — the development and construction of indicators — went ahead. During the period 1990–1991, two technical groups examined a range of issues concerning definitions and comparability, and sought to iron out some of the technical problems involving the organisation of data gathering and processing. A new Network dealing with issues of education and labour market destinations was created under the leadership of Sweden. The CERI Governing Board now stood solidly behind the Indicators project. It instructed the various Networks, technical groups, and the secretariat, to draw up a handbook to provide a clearer account of the conceptual and organisational framework for data collection and management, and led to a set of refined international education indicators. It also stressed the need to disseminate information about indicators, which created a climate of support among policy makers and analysts across member countries and even beyond. A Consultative Group to ensure greater coherence across the project as a whole was also established.

In September 1991, the second phase of the INES project culminated in a major meeting at Lugano where the first draft edition of *Education at a Glance* was presented. It contained data on thirty indicators which ranged from relatively traditional items such as participation rates, to complex and contested measures such as characteristics of decision-making within education systems. The meeting also launched *Making Education Count* (CERI, 1994b), which addressed a range of conceptual issues. It revealed how far matters of definition, bias and validity of comparison and inference remained unresolved. How comparative data should be applied in measuring the relative progress a system might have made against particular objectives remained unclear, as did the question of relative weighting a system might attach to specific indicators within its framework of priorities.

Nevertheless, the basic principles underlying the INES work had been well and truly established, though further conceptual and technical work was required if the consolidation and refinement of indicators was to be ensured and if the range of countries and organisations involved were to be extended. Phase Three consolidated the work of the preceding phases. The OECD viewed this as a period during which a wider range of indicators would be produced, and an organisational framework for sustained work on indicators would be tested and established. During this period, further conceptual work was undertaken, methods and techniques were improved, and attempts were made to show the ways in which the indicators could serve in various ways related to policy.

Organisationally, a Policy Review and Advisory Group was established, working closely with the secretariat, to establish a stronger anchor for the work in policy. A group of national coordinators was appointed to consult with the secretariat in the oversight of the operational aspects of INES, and to "contribute to the diffusion of an indicator culture within education circles" (OECD/CERI, 1995a, p. 4). Also significant was the integration of the various statistical activities in education under a new division within DEELSA, the Statistics and Indicators Division. The thinking behind this move was to put together the resources of the Education Division and CERI, but more significantly and symbolically, the move represented the mainstreaming of the OECD's indicators work in education, from developmental status to being a core activity of the Organisation. The project was put under the direction of a 'new guard' of statisticians, finalising its move away from its philosophical starting point.

The success of the project meant that resources were increased to improve and expand the indicators, though the major part of the resourcing came from member countries themselves through the funding of the networks, each sustaining at least three or four full-time staff (IT 17: secretariat, 1997). In 1996, *Education at a Glance* was accompanied by a shorter analytical volume, *Education at a Glance – Analysis*, which discussed "some of the key themes emerging from the data and the lessons the results carry for education systems" (CERI, 1996b). Subsequently, the *Analysis* was released at a different time from *Education at a Glance*, reflecting "the continuing development of two distinctive but highly complementary publications" (CERI, 1998b, p. 6). In 1997, the secretariat was requested by OECD Council to begin preliminary work on comparative indicators of human capital investment. The exercise, it was thought, would:

" . . .   be the first step in developing an international set of indicators to answer the interests of policy makers and, I suppose, people in education and employment ministries. People are under pressure to justify their expenditure so they want data to show that education and training outpays investments rather than consumption activities."

(IT 8: secretariat, 1997)

Amongst the dimensions considered were "trying to value the impact of enterprise training on earnings and job tenure of individuals, trying to possibly estimate the stock of human capital or the level of human capital, and also how that impacts on long-term economic growth" (IT 8: secretariat, 1997).

Interest in education indicators is not restricted to the OECD and its member countries. Other inter-governmental organisations — UNESCO and the Asia–Pacific Economic Cooperation forum (APEC) — have pursued similar agendas. Indeed, the OECD and UNESCO's work on indicators furnished a precedent for APEC's interest in developing indicators of school effectiveness (APEC, 1997, p. 4). In 1995 UNESCO, OECD and EUROSTAT (the statistical wing of the EU) joined forces to collect data on key aspects of education, thus consolidating a liaison first created when the OECD adapted the International Standard Classification of Education (ISCED) Systems originally developed by UNESCO, and which in turn was based on the OECD's earlier development work (Papadopoulos, 1994, pp. 53–54). Collaboration, though fraught with difficulties (IT 19: bureaucrat, 1997), explored common definitions, use of criteria for quality control and improved data documentation in order to improve the international comparisons of education statistics. Reflecting this expanded terrain, the 1998 edition of *Education at a Glance* included data from a wide range of non-member countries gathered through the 'World Education Indicators Programme' (WEI) conducted in collaboration with UNESCO and partially funded by the World Bank. World Indicators pinpointed differing outcomes between OECD and WEI countries in matters such as student demography, levels of educational attainment, graduation rates and resourcing per student (CERI, 1998a, pp. 29–30). By 1998, in the OECD's own words, indicators were covering, "almost two-thirds of the world population" (CERI, 1998a, p. 6).

In short, the 1990s saw some remarkable shifts in the development of educational indicators within the OECD: from philosophical doubt to statistical confidence; from covering some countries to covering most of the world; from a focus on inputs to a focus on outputs; and from occupying an experimental status to being a central part of the Organisation's educational work. What purposes, are indicators seen as serving?

Indicators For What Purpose?

This brief history of the OECD's work on education indicators should be sufficient to show that *Education at a Glance* is a product of much deliberation and disagreement over its characteristics and purposes, both within the organisation and elsewhere. What has emerged is a model of indicators both sophisticated and distinctive. The model is located historically within the OECD's task of indicators research. It is informed by an acute awareness of education's central role in individual and economic development, and by a particular view of the contribution that comparative indicators can make to policy development and administrative reform. In this, the OECD's model of education indicators differs from the indicators commonly discussed in the literature.

Oakes (1986) has pointed out that the broad purposes of indicator systems are "to measure the health and effectiveness of the education system and help policy makers make better decisions" (cited in Nuttall, 1992, p. 15). According to Nuttall, "the allusion to the medical check-up is appropriate since it draws attention to the skill of the practitioner in interpreting and making sense of the information available from many sources ...   — information that can help in diagnosis but is not prescriptive or judgemental" (Nuttall, 1992, pp. 15–16).

In an earlier study of social indicators, Carley (1980) described four different types of indicators: informative indicators, predictive indicators, problem-oriented indicators and program evaluation indicators. Informative indicators describe the social system and the changes taking place within it. These are indicators produced at regular intervals around a range of relevant variables. Predictive indicators describe statistical relations between environmental variables which attempt to predict trends. Problem-oriented indicators point towards policy situations and actions on specific social problems. And program evaluation indicators operationalise policy goals the better to monitor the progress and effectiveness of particular policies and programs. In more recent years, performance indicators have been used widely by systems and organisations to meet demands of accountability. In effect, they are measures that describe how well a program meets its objectives.

The OECD's education indicators do not fall into any of the four types identified by Carley. They contain elements of each, but also other more distinctive characteristics. Nor do they fit Oakes' definition because of their evaluative component. For a long time, the OECD's researchers exercised caution about cross-national comparisons, largely because the OECD's approach operates by consensus. This consensual style, intended to provide general information that is relevant and useful to each of its members, makes determining a basis for comparison difficult. Within the organisation, as Bottani and Walberg explained, even relatively simple indicators such as student enrolments cannot be taken lightly: "Sometimes goals and cultures are so specific to each country to cause difficulties of measurement and comparison" (Bottani and Walberg, 1992, p. 7). According to the OECD, interpretation is always a complex exercise that involves referring to a range of factors, of which the indicators data is just one. They are to be seen as an aid to judgement. As Bottani (1995, p. 11) described, indicators "provide decision makers and users of the education system with information on the results obtained, the way the system operates, the effectiveness of the service provided, the problems involved and the outlook." In short, indicators help 'to render these systems more transparent'.

Despite sensitivities to the problem of interpretation, the OECD has been re-lentless in developing a system of education indicators involved not only with the description of systems but also their evaluation. The Indicators project presents a comparative assessment of the educational processes and outcomes of the policies and programs pursued by the OECD's member countries. Key policy questions ask not only "how many?" and "how much?" but also "how good?" (Nuttall, 1992, p. 13). Education indicators are conceived to provide information that de-scribes a system's progress in achieving certain desired educational outcomes, or highlights features of the system known to be linked with desired outcomes. The purpose is to measure education's current effectiveness, or at least provide infor-mation of help to policy makers in making course corrections to predict future performance:

> "A quantitative description of the functioning of education systems allows countries to see themselves in the light of other countries' performance. Through international comparison, countries may come to recognise weaknesses in their own education systems, while also identifying strengths that can otherwise be ignored in the heat of domestic debate."

> (CERI, 1998a, p. 5)

In this respect, OECD indicators differ radically from programmatic indicators which respond largely to the particular information needs of various agencies, and are therefore localised. In contrast, the OECD education indicators are comparative, and are developed against a set of agreed criteria that constitute the reference points against which performance is measured and judged.

The Indicators project attempts to reconcile contradictory purposes: on the one hand, a long tradition of cross-country comparison which takes account of the particularities of national histories and circumstances and the inherent limitations of such comparison; and on the other hand, comparison for the purposes of evaluation against common criteria. The Organisation claims that education indicators provide an opportunity for countries "to learn more about themselves" and cautions that "the science of understanding and interpreting international education indicators is still in its infancy" (CERI, 1996b, pp. 9–10). As such, it warns, indicators should not be viewed as providing international league tables of performance, as dictating particular benchmarks for performance, or as predicting the outcomes of particular policies. It maintains, for instance, that:

> "Numbers can never on their own give an adequate understanding of trends in education. Quantitative indicators are often a necessary condition for sound analysis of educational developments but they are never sufficient. For example, tests of student achievement can yield interesting information in comparing the performance of education systems, but without an understanding of the curricular objectives of each country it is impossible to evaluate the success of their respective systems."

> (CERI, 1996b, p. 10)

However, the very nature of the project assumes a commonality of structures, processes and priorities as a basis for the development of indicator systems. As Alan Ruby (1992, p. 77), former chair of the Education Committee, noted, the OECD education indicator system makes the "assumption that schools, the educational process itself and school or educational systems are essentially comparable". Thus the significance of the comparability element in the project was emphasised in the first edition of *Education at a Glance*:

> "Apart from data on teacher supply and student enrolment, there was until recently relatively little demand ... for data enabling international comparisons to be made. However, the mid-1980s witnessed a major change in attitude in this respect; and international comparisons are now widely seen as a means of adding perspective to the analysis of education systems at a national level."

> " ... the indicators are selective and intended to be policy relevant, providing information useful for decision-making and evaluation ... [and] in addition to being reliable and valid at the national level, the indicators are standardised in a way that makes them comparable among the OECD countries."

> (CERI, 1992, pp. 9–10)

By 1996, the OECD took the view that it was "possible to produce a limited set of up-to-date and internationally comparable indicators on education" (CERI, 1996a, p. 11). The 1996 edition of *Education at a Glance* asserted that the indicators "represent the consensus of professional thinking on how best to measure the current state of education internationally, tempered by the availability of valid, reliable and comparable data" (CERI, 1996a, p. 9). Two years later, it was claimed once again that "the continuing implementation of common definitions, the use of high standards for quality control, and better data documentation have improved the international comparability of education statistics" (CERI, 1998a, p. 9).

Despite the considerable range of problems associated with gathering meaningful comparative data, this point is worth further elaboration, if only to pose the question later as to why the Organisation and its members remain deeply committed to a project so fraught with pitfalls.

The Problem of Comparability

The OECD does not itself collect the data upon which its indicators are based. Its role is in the selection and compilation of data already available. There are two major sources: data supplied by member countries and data derived from large scale international surveys, for instance, the International Adult Literacy Survey. Drawing from these diverse sources, the OECD refines the data, clarifying definitions, aligning them to comparable categories and determining the links between input, processes and outcomes. The technical and political difficulties associated with this exercise are considerable. Over the time that *Education at a Glance* has been produced, the coverage of countries supplying national data has varied. For the earlier volumes, some did not supply data.

It was not until 1995 that Korea and Mexico joined the education indicators program. Aligning data supplied by member countries has proved notoriously difficult. National data can often be incomplete, unreliable and out of phase in terms of timing and methods of data collection. The information supplied by federal states like the US, Australia, Canada and Germany provide data in terms of weighted means, a process that cannot be assumed to have been carried out in any uniform fashion. Even aggregations are not always reliable because of changes in definitions and methodology. This is particularly so in collecting data on participation in tertiary education, where reforms in the post-secondary sector often change the ways students are classified for the purposes of allocating grants and benefits.

Data derived from international surveys suffer from similar problems. For example, national policy changes seriously affect the classification of national educational programs according to the levels defined in the International Standard Classification of Education. In 1996, *Education at a Glance* included indicators on student achievement in Mathematics. Four indicators on achievement of thirteen-year-olds are generated from the Third International Mathematics and Scientific Study (TIMSS), conducted during 1993–1994 by the International Association for Educational Evaluation. This study measured student performance in mathematics and assessed factors contributing to this performance. Whether this data source is reliable is an issue about which the OECD itself does not appear confident. Normal sampling errors apart, the 1996 edition of *Education at a Glance* acknowledged the possibility of a number of non-sampling errors in TIMSS due to differences in the ways the survey was carried out in different countries. It admitted that how far this introduces bias in the results "cannot be easily determined" (p. 197).

The OECD readily acknowledges many deficiencies in the current system, deficiencies which relate to coverage, validity, comparability, accuracy and timeliness of the indicators. It insists, however, that these deficiencies are being overcome thanks to further methodological refinements. It is, therefore, cautious about claims made on behalf of the indicators. It recognises that "The diversity of education systems and differences in the structure of governance of education provide a challenge for international comparisons" and lists no less than nine significant problems of comparability (CERI, 1996a, p. 12). Most crucial among these is the problem relating to the classification categories within ISCED which, the OECD argues, "in its present form, is ill adapted to current needs and imposes limitations on international comparisons of education statistics as well as the analytical use and interpretation of the indicators" (*ibid.*). Other stumbling blocks arise with double-counting, inconsistent classification of types of educational programs, those associated with the classification of educational funding, inadequate methods of estimating full-time participation, and ambiguities involved in assessing graduate outcomes. (See also Steedman, 1999.)

The problem of comparability lies at the heart of the Indicators project, especially under conditions of decentralisation. In a decentralised system where individual nations, states or other units have autonomy and responsibility for developing their own statistical systems, data collection procedures used by each of these supplying data will have evolved distinct definitions and categories to meet specific needs. The key terms contained in the survey instruments may indeed refer to features specific to the local context, for example, the term 'college' has a variety of different meanings. This being so, then how may comparability of statistics collected in decentralised systems be ensured, let alone improved? Selden (1992, pp. 107–113) described a project in the US that sought to resolve some of these problems. In essence, Selden's solution is to recommend standard definitions and procedures, to develop stricter guidelines for reporting statistical information and to monitor adherence to standard definitions and procedures in the collection of data.

Certainly, this approach might work in a country where central government retains a great deal of residual power. But, it is unlikely to be effective where local education units resist and even refuse to accept the central authority. Moreover, Selden's solution may be regarded as an attempt to undermine the principles of decentralisation. Either way, since the OECD lacks any significant legislative authority over member countries, it is not in a position to pursue such a tactic. Two courses of action alone remain open to it. First, it can statistically moderate the data supplied and work with various techniques of aggregated means and approximation. This the Organisation does. The second course of action is for the OECD to use its powers of suasion to encourage its members to collect data in line with the statistical requirements of its International Education Indicators System. This might include encouragement to adopt the OECD's definitions and procedures, and to collect data on nominated indicators, if this is not already done.

Some evidence suggests that the OECD has been successful in reshaping the statistical systems of its member countries in this way. Denmark, France and Iceland have begun to collect and supply data consistent with the model developed by INES. Canada decided to develop a set of pan-Canadian education indicators on the basis of INES experiences (OECD/CERI, 1995a, p. 4). Whether the data collection requirements have also influenced member countries to rearrange their policy priorities, however indirectly, can at present only be speculated upon. Raffe's research (Raffe, 1998a, pp. 594–595), for example, on the UK's take-up of comparative data to inform vocational education and training policy found that "the use of comparisons for policy learning was informal, haphazard and ad hoc". By contrast, Germany's 'disappointing' results in the TIMSS studies lay at the heart of its decision to participate in a further OECD comparative exercise to allow comparisons to be made between the *Laender* — a remarkable step, according to Weiss and Weishaupt (1999), in light of "the strong reservations of German teachers and administrators about standardized quantitative approaches to evaluating the performance of schools and school systems" (1999, p. 117).

Regardless of how comparative data in fact feed policy debates in member countries, the very process of drawing in an expanding number of countries into a single comparative field is significant in itself. Inevitably, the establishment of a single playing field sets the stage for constructing league tables, whatever the somewhat disingenuous claims to the contrary. Visually, tables or figures of comparative performance against an OECD or country mean carry normative overtones, as do more recent comparisons between OECD and non-OECD countries in the World Education Indicators programme. To be above, below or at par with the OECD average invites simplistic or politically motivated comment, despite the pages of methodological and interpretative cautions which abound in the annexes of *Education at a Glance*.

Illustrations of such comment can be found in the collection of media clippings the OECD collates to monitor reception to its annual publication. In the collection of April 1995 (OECD, 1995d), more than half of the reports indulge precisely in the crude country comparisons, feared by the OECD. Headlines trumpet simplicities along the lines of "UK weak in nursery education", "UK spending better than many", "Scots go right to the top of class", "Ireland neglects primary education", "Irish pupils score below average", "Teaching time is the highest in USA" and "Mixed marks for NZ in schools report". The texts beneath these headlines are invariably highly judgemental, preventing any possibility of a dispassionate understanding of the issues. Often, far from 'dispelling myths', indicators serve to legitimise preconceived notions of educational performance and problems, a situation as true among policy players as it is in media debates about education. In Australia, for example, the same set of indicators have been used both by governments to suggest relatively high levels of public funding for education and by teacher unions (IT 14: secretariat, 1996) to uphold the charge of poor levels of expenditure.

Despite problems of comparability and oversimplification, the OECD remains committed to its Indicators project and the project remains well supported. Why should there be such a strong level of support? What is the project's broader significance? To answer such questions requires that we return to a more theoretical terrain. The analysis of performativity raised in the previous chapter will be developed further and placed in relation to two related but independent dynamics: the first turns around technical rationality as an expression of cultural globalisation; the second involves the neo-liberal revival of human capital theory associated with economic globalisation.

Indicators and Global Rationalisation

In discussing technical rationality, we will begin with Loya and Boli's (1999) case study of global standards organisations which explores how processes of standardisation (e.g. of weights and measures) contributed to "the construction of a uniform built environment" (p. 169). This study draws on Meyer et al,'s (1997) world polity theory which at one level and simply stated, extended Weberian notions of bureaucratic technical rationality to the global stage. Naturally, there are limits to the parallels between the development of international educational indicators and the thrust for, say, standardisation of credit card thickness, film sensitivity or the symbols used on automobile dashboard controls. Nevertheless, some of the assumptions behind such exercises of standardisation may serve to explain why educational indicators are so well supported.

Standardisation works through an extensive network of national, regional and global standardisation organisations. The authority to coordinate activities is "increasingly lodged in the global-level organisations, regardless of the individual contributions or influence of lower-level entities" (Loya and Boli, 1999, p. 176). While a great deal of variation exists at local and national levels, there is remarkable similarity of these organisations. As Loya and Boli pointed out "National bodies are compelled to be virtually isomorphic with each other as a prerequisite for membership in global bodies" (p. 177). As with the Indicators project, international standards organisations are sustained by networks of technical committees whose members are required to set aside the particular interests of their own countries the better to pursue the common goal of standardisation. The committees work democratically. At the international level, each country regardless of size has one vote. As with the Indicators networks, so within international standards organisations, member agencies bear the bulk of the costs associated with standardisation, and much of the work is undertaken voluntarily. International standards organisations operate on a consensus basis. Powers of suasion are used rather than coercion:

> "The preferred means of *global* standardization is universal voluntary compliance, a principle that thoroughly infuses the standardization sector. The logic of the sector is quite simple: the character of the standards and the nature of the processes by which they are generated — universal, consensually derived standards of unimpeachable technical merit — are themselves sufficient rationale for their adoption. Add to that the compulsion of market competition, and coercive mechanisms are entirely unnecessary in the increasingly integrated world economy. As the BSI [British Standards Institute] notes, 'pressure to conform has never been stronger'."
>
> (Loya and Boli, 1999, p. 181, original emphasis)

The Indicators project evokes similar comments:

> "Because they [the networks] work on a consensus basis, they try to avoid conflict or trade-offs. 'I'll support your indicator if you support my indicator.' It's never posed like that. Rather, it's a give–take. And you can do this as long as you don't go too far in making those positions too hard — too explicit. Of course countries in a gentle kind of way indicate where their priorities are."
>
> (IT 17: secretariat, 1997)

> "In the INES project, there is a positive consensus and agreement between all countries concerned so that, for example, Luxembourg has the same voice as that of the United States, but it doesn't carry the same weight in terms of funding or political influence obviously, so it's true that some countries, because of their sheer size and importance in terms of financing the INES project ... will carry more weight in terms of their view. That doesn't necessarily mean that they set the agenda.

I think there is always a balance struck ... So I would say that certainly no country dominates the agenda, some countries have more influence."

(IT 8: secretariat, 1997)

As Loya and Boli observed in relation to standards organisations, it is not that they are free of conflict, simply that their structure: "is designed to activate transcendent purposes oriented around global progress. Helping promote such commitment among individuals, firms, and associations is itself a central purpose of these organizations" (p. 183). It is not without significance that the underpinning rationale of standardisation is not technical. Rather, Loya and Boli suggest, the BSI sees its goals as 'emphatically social', leading to such benefits as improved health and safety protection, reduction of waste, enhanced product quality and reduced prices (p. 181). "Indeed, standardization is routinely portrayed as the rational means to solve some of humanity's most serious problems ... A pragmatic utopia, not just an inchoate world of technical marvels, is the implicit ultimate vision" (Loya and Boli, 1999, pp. 181–82). In general, Loya and Boli conclude, global standardisation can be seen as a "highly institutionalized sphere of world-level collective action" (p. 191) — non-partisan in nature, targeting all of humanity as its beneficiary:

" ... the global standards project is highly rationalized, both in the social theory that motivates actions (standardization contributes to a broad range of desirable human purposes) and in the procedures employed ... In short ... standardization organizations are built on world-cultural conceptions of universalism, rational progress and egalitarianism."

(p. 192)

This is Enlightenment writ large and as such constitutes a powerful motivation for participation. That educational indicators are not of 'unimpeachable technical merit' is widely recognised. But the Indicators project *is* underpinned by an assumption of global progress, in terms of its ever-expanding reach and in its ontological rationality. Ruby, for example, observed: "We invest in the OECD's education work because it is likely to inform and improve our policy making. I appreciate that this is a highly rational view of the world based in a belief that information will improve things" (Ruby, 1997, p. 10). The 'we' in that context referred to Australia, but the broad thrust of his presentation concerned policy directions in Western nations generally. In one sense the constant efforts of refinement and improvement implicit in the Indicators project may be interpreted as a quest for perfectibility — in short, a pragmatic utopia.

Rationality, however, represents more than a value stance. It also entails a particular — positivistic, data-driven — policy approach. The commitment to rational policy making "underscores the high priority the education authorities give data collection and comparability of data" (Ruby, 1997, p. 6). The demand for data-based information from all OECD countries, Ruby argued, was driven by the political imperatives of accountability:

"Rational models of decision making reinforce this trend [towards accountability]. Management by objectives, program budgeting and the like all depend on the explicit goal setting and documentation generating more information and a demand for existing information to be transformed into new reporting frameworks."

(Ruby, 1997, p. 8)

Within Ruby's general argument, two strands of rationality converge. One was associated with the French republican ideal which formed one guiding force behind the Indicators project. The other drew from US-inspired new public management theory, the organisational accomplice to neo-liberal economic theory. Related to this second strand, the management, measurement and comparison of performance — of individuals, systems and nations — within a global market economy have become central components in the new forms of educational governance which goes hand in hand with the new policy consensus outlined earlier in Chapter 2.

The Comparability Imperative and the New Educational Policy Consensus

The new consensus in educational policy that accompanies globalisation, of which the OECD has been a major advocate, is held together by a micro-economic focussed human capital theory that is unlike earlier forms of macro human capital theory. While the latter argued rather crudely for a supposed link between levels of educational expenditure (inputs) and economic growth and competitiveness (outputs), the new micro approach in contrast emphasises the specific skills of individuals thought necessary to participate effectively in a knowledge based global economy. From a policy perspective, education and the provision of multiskilled individuals who are flexible and adaptive to rapid change and uncertainty have become central elements of economic policy in all OECD countries.

Together with this reworked version of human capital theory went the advocacy of new forms of governance dedicated in a post-Keynesian sense to achieving better educational outcomes (more effectiveness) at lower costs (more efficiency). The structural or organisational profile is less hierarchical with tighter steering over a narrower policy agenda wielded by a strategically focussed centre. In OECD member such structural changes emerged in various ways depending on local politics and political cultures. Yet, restructuring of education systems calls apparently for tighter steering with greater differentiation — sometimes referred to as 'tight–loose' forms of governance — particularly in member countries where a neo-liberal marketised approach to education predominated. Performance indicators, as we saw earlier, serve as central mechanisms of accountability in such tight–loose forms of steering and become a means whereby the centre may apply 'course corrections' over policy directions. The Indicators project feeds into and extends the paradigm of accountability by creating a global stage for comparison — a condition reflected in the comment quoted at the start of this chapter, namely, "comparability now belongs with accountability to that changing set of driving words which shape the current management paradigm of education" (Alexander, 1994, p. 17).

From a normative standpoint as seen in recent attempts to measure outcomes and return on investment in particular forms of education, indicators are both reflective and expressive of the new micro-economic approach to human capital theory and of the associated culture of performativity. Here, the introduction to the second edition of *Education at a Glance* is suitably revealing:

> "This second set of education indicators appears at a time when OECD Member countries face serious problems of sluggish growth and rising unemployment ... "

> "The OECD countries must continue to search for ways to increase the quality, equity and efficiency of their education systems. The importance of this role can be inferred from the economic role of education. Thus, while the social and cultural value of education remains fundamental, today's knowledge intensive societies call for a growing emphasis on education as a key investment."

> (CERI, 1993, p. 9)

Again, the third edition of *Education at a Glance* (OECD, 1995d) summoned up a context where "the relationship between employment and education is to a large extent global" (p. 7). Several indicators were devoted to exploring "the close relationship between education and the labour market". With a similar emphasis, the opening paragraph of the 1996 volume of *Education at a Glance* commented:

> "Throughout OECD countries, governments are seeking effective policies for enhancing economic productivity through education, employing incentives to promote the efficiency of the administration of schooling, and searching for additional resources to meet increasing demands for education."

> (CERI, 1996a, p. 9)

Two years on, the cover of the 1998 edition of *Education at a Glance* proclaimed its achievements in providing "a richer, more comparable and up to date array of indicators than ever before" — including taking "a further step to reporting internationally comparable data on lifelong learning and its impact on society and the economy."

These statements clearly endorse an economistic view of education; that education is best viewed as an investment to help foster economic growth. As a matter of course, the OECD recognises that education also contributes to the goals of personal and social development; the policy conclusions of the 1998 analytical volume are by no means an uncompromising economistic reading of education. In the sphere of lifelong learning, the continuing importance of a "predominant public stake in financing tertiary education" is stressed (CERI, 1996b, p. 6); in "giving attention to the *processes* of teaching and learning", and not merely inputs and outputs 'in terms of measurable student achievement' is upheld (p. 38, original emphasis); in the area of youth pathways, the coherence and structured approach of the Nordic countries is praised in contrast to a more *laissez faire* approach (p. 55), and when turning to funding tertiary education, attention is drawn to the problems of 'user pays' systems excluding students from low income families (p. 72). The social dimensions of human capital also receive recognition from within the secretariat:

> "In reality human capital is quite a wide-ranging concept and it involves also social capital, what we call organisational learning which resides in institutions and organisations, and it's not just the job-related skills [or qualifications] of individuals ... because they can be quite a poor proxy to the whole gamut or the whole range of skills which people have. So it is a difficult conception to be tied down and clearly we do have to restrict ourselves to a more operational and shall we say policy-malleable sort of idea of human capital and that means we certainly have to look at it from a labour market standpoint."
>
> (IT 8: secretariat, 1997)

By focussing on the 'operational' and 'policy-malleable', whatever the concessions made to the complexity of the concept, human capital becomes framed and set, overall, in more straight-forward terms — and harnessed to the interest of national (and global) economies. Such a focus helps explain the strong support for the Indicators project from member countries. Indeed, the opening paragraph of the 1998 analytical volume underlined this constraint:

> "National economies are restructuring themselves in ways that react to technological, social and economic change, and at best take advantage of them. A universal objective has been to give greater weight to the skills, knowledge and dispositions embodied in individuals. The value given to such human attributes, together with a continued rise in levels of education, income and wealth, drive increased demand for learning in its broadest sense. Education and training systems, institutions, schools and programmes are being asked to respond to higher expectations and they must do so under very tight budgetary conditions and keen competition for public and private resources."
>
> (OECD, 1998b, p. 5)

There is, then, some mismatch between the broadly framed policy conclusions noted above and the core framing of the issues. Lifelong learning is framed in terms of constant skills renewals and teaching is viewed in terms of motivating and preparing individuals for a lifetime of learning (p. 6). Learning itself is presented on the one hand in terms of transitions and pathways through education and on the other between education and employment. Tertiary education is deemed to contribute to lifelong learning, so that "all learners in tertiary education might be expected to contribute a share of the costs of their tuition" (p. 6). In short, from a normative stance, an effective education system is held to be an efficient system expressed in terms of returns on investments made and as individual and labour market outcomes of education rather than concerned with the social, ethical and cultural dimensions of education. By drawing all countries into the same comparative field, the Indicators project reposes on the assumption that all systems, 'in their own ways' will/should strive for similar 'effective' ends.

In a similar spirit a parallel move seeks to orient the indicators to measures of outcomes. The introduction to the 1996 *Analysis* remarked: "Finding key international indicators of educational outcomes" while "a difficult task" is necessary for the measurement of the effectiveness of systems (CERI, 1996b, p. 8). It suggested that two sorts of outcome measures were required: first, international benchmarks of performance for core school subjects in reading, mathematics and science; second, tests of adult competencies. Having taken such a stance, the *Analysis* argued: "Both kinds of tests need to be carried out on a more regular and systematic basis to meet the needs of policy makers" (CERI, 1996b, p. 8). While recognising that the establishment of such tests is not cheap for governments, the *Analysis* asserted that testing is "essential for any meaningful understanding for the effectiveness of education systems in an international context" (p. 8).

Such an interpretation moved the Indicators' project beyond merely the task of helping countries identify strengths and weaknesses: it promotes a 'corrective policy'. Hence, the statement about the purposes attributed to the framework for lifelong learning takes on very specific significance and overtones:

> "For the first time, a *monitoring tool* is advanced which can be used to take stock of the present state of play and to track progress toward the realisation of lifelong learning for all. The framework ... is intended to make more specific the links between aims, policies, practices and results, and to *overcome the drift* in the policy debate."
>
> (CERI, 1998b, p. 5, our emphasis)

Given the OECD's mode of operating, the effectiveness of monitoring processes remains a moot point. However, that is not the issue. Rather, it is our argument that, regardless of how indicators are used or abused, a 'corrective tendency' is in-built in their very extensiveness and in their embodiment of the logic of performativity itself redolent of new forms of educational governance. Such a logic thrusts forward a degree of policy homogenisation which exists over and above more distinctive policy orientations at the national level.

Conclusion: Indicators and New Circuits of Homogenisation and Differentiation

When Weber explored the processes of rationalisation which accompanied modernity, he concentrated upon those structures within the bureaucratic arrangements of nation-states. In the perspective developed here, performative rationality in association with new forms of governance has displaced these processes to another, global, level of development. The OECD served as an important bearer in this shift acting — to use the words of the Ministerial Council

Communique of 1997 — as "an especially potent instrument of global change and reform" (OECD, 1997f, p. 24). The direction of that change and reform rested on the credo that within a global economy the competitive advantage of nations depends on the quality of its human capital (Reich, 1991). This was a view emphasised by the Director of DEELSA at the Lugano meeting in 1991 which marked the end of the second phase of the INES project:

> "Education and training appear more and more as one of the key factors likely to influence productivity and competitiveness in the long term. While this is in line with some of the classical thinking on the economic returns of educational investment and the theory of human capital, it is a theme now imbued with a social and global urgency it did not possess before."
>
> (Alexander, 1994, p. 16)

Global urgency, it would seem, stimulated a keen interest in human capital investment; it set out the notion of a globally competitive playing field. The 1995 volume of *Education at a Glance* in its introduction proclaimed: " . . . as global influences have greater impact on societies and economic activity, educational performance has to be considered at a level beyond the traditional national context" (CERI, 1995b, p. 7). As a member of the secretariat remarked:

> " . . . there is a close connection between the interest in international comparisons on the one hand and intense global competitiveness and pressure of that sort, and it's a connection which I think comes more readily to people than it used to."
>
> (IT 10: secretariat, 1997)

Earlier, it was argued that both cultural homogenisation and heterogenisation are inherent in the flows of globalisation. From this perspective, the Indicators project contributes to the drive towards cultural homogenisation. From what we have seen, the lessons for policy which the OECD draws from their *Analysis* documents also point out directions for member and non-member countries to follow. Here indeed the OECD acts as an international mediator of knowledge rather than being confined to serving as a catalyst for change. In this way, the Indicators project forms an integral part of the OECD's broader task of merchandising a particular thesis of educational purposes (new human capital theory) and of governance (new public management theory). These two items exist in tandem; new public management may be regarded as the organisational template of the neo-liberal state.

Weber associated rationalisation with disenchantment of the world and both accompanied by loss of meaning. The new rationality embedded in the Indicators project also evokes a sense of loss for some in the secretariat who hold what has become a minority viewpoint:

> " . . . the enthusiasm of the statisticians has been to produce more and more separate measures, and then of course you try in some curious way to put them together through technical means. The basic problem is not that. The basic problem is a

conceptual analysis of what we mean by 'functioning education system', and not what are the 500 features of it ... I'm somewhat isolated in this view."

(IT 56: secretariat, 1995)

The 'technicisation' of education is not the only aspect of this intellectual deprivation. Meaning becomes less nuanced, an equal loss, as elements of national systems become decontextualised or evacuated from their local setting and are recontextualised into more global comparative frameworks. In this transmutation, local differentiation is at odds with a broader homogenising imperative. Loya and Boli identify a similar phenomenon in relation to standardisation:

"The uniformity engendered by standardization is deep and far-reaching, but it is also subtle. It reduces fundamental differences and provokes the intensified reification of superficial differences: varied facades attract much attention, but underneath they are hardly distinguishable."

(p. 197)

Applied to education, this condition raises important implications for the meaning of national policy making. Finding expressions of national differences is not difficult. The 1998 *Analysis* (CERI, 1998b, pp. 11–12) set out the main dimension within various national approaches to lifelong learning. The table reveals how far the rhetoric of member countries is inflected by nationally or regionally distinctive ideological and political currents. Australia, for example, focussed on 'skills and knowledge enhancing job chances and personal enrichment'. The EU qua OECD member, stressed the 'role of education in constructing active European citizenship, recognising different cultural and economic approaches but also the commonality of European civilisation'. The EU's objectives, apart from promoting various skills, included 'fighting exclusion and equal treatment of human capital and other forms of investment'. Finland, likewise, underlined democratic values, social cohesion and internationalism, productivity and competitiveness as well. The Netherlands emphasised social and economic justification for lifelong learning. For Japan, amongst other things, 'lifelong learning aims to remedy problems arising from the pressures of a "diploma society" '. The UK and the US alike put weight on literacy standards. The former sought, additionally, to develop 'the spiritual side' of individuals — surely a post-Blair sentiment. The latter aimed to connect every school and library to the Internet by 2000.

Against this background the Indicators project constitutes one form through which the globalisation of education policy is expressed and through which it expresses itself. Since globalisation both homogenises and differentiates, the question of how far the homogenising force of the Indicators project may be offset by local priorities, remains. There is no clear answer, though one should note that global influences do not work deterministically to override the capacity for national policy making. Let us continue with lifelong learning as illustration. Given the diversity of interpretations found in this concept, the OECD's struggle to evolve a commonality for the basis of comparable indicators reduces almost of necessity, on the one hand, to terms of high abstraction, *pace* the development of a "culture of learning" (CERI, 1998b, p. 13), and on the other hand, retreats into measurable characteristics, as for instance adult participation in education (p. 15). The slippage, then, between local practices, national policy rhetoric and recontextualised global statements is vast. Nevertheless, the OECD clings to its vision of a pragmatic utopia:

> "Monitoring progress will be a difficult and always imperfect exercise: it will never be possible to construct fully adequate indicators of all the informal learning that occurs in people's lives. But it will nevertheless be possible to build on our present understanding as an aid to policy development. Already it is possible to show that only a minority of the OECD population is participating in education and training on a lifelong basis. There is a considerable distance to go in making learning a reality 'for all', even without considerations of context, quality and relevance. Attaining the goal would be costly but it is also an investment. It is a realisable ambition, if it is pursued as a long-term effort to which all partners contribute."
>
> (1998b, p. 23)

The broader point that may be marked is that support offered to the Indicators project would point to the OECD's effectiveness in implanting its desired "indicator culture within educational circles" (OECD/CERI, 1995b, p. 4). Among both bureaucrats at the service of globalisation and policy elites, an epistemological consensus appears to exist. As Rose (1999) noted, this consensus turns around a new and powerful paradigm of policy as numbers.

6

From Recurrent Education to Lifelong Learning: The Vocational Education and Training Saga

"We should break the monopoly of the 16–19 age group on access to higher education — it will always favour the children of the existing social elites; spread the *right* to education to the adult, and in particular the early adult years; encourage firms, trade unions and public administration to accept responsibility for developing individuals through education and training; allow more flexible procedures for acquiring professional qualifications . . . "

(Ron Gass, Director of CERI, cited in Duke, 1974, p. 8)

"Success in realising lifelong learning — from early childhood education to active learning in retirement — will be an important factor in promoting employment, economic development, democracy and social cohesion in the years ahead."

(OECD, 1996d, p. 13)

How does OECD work intersect with the work of nation states? What are the universalising tendencies within certain key policy agendas? Some of the ideas raised in earlier chapters about how the OECD exerts influence, and how OECD's agendas themselves are shaped by globalisation are developed further. This chapter sets out to show how the OECD may be used as a forum and policy instrument. It tracks how the OECD, as an 'international mediator of knowledge' fits as a policy actor in opening up policy to further convergence. It discusses in more theoretical terms the meaning of *national* policy making when it is confronted with the globalisation of educational policies. These issues will be examined by focussing on vocational education and training policy in recent years. Particular reference will be made to developments in Australia and within the OECD. This theme was chosen in recognition of Dale's point that given their explicit links with the economy some domains in education are more highly charged with respect to globalisation than others (Dale, 1999).

For Australia, vocational education policy has particular salience in fostering long-term links with the OECD. In 1974, the Commonwealth "pointed the way" (Papadopoulos, 1994, p. 148) by being the first country to commission an OECD country review on "the transition from school to work". It commissioned a second country review on the related topic of youth policy in 1977. In 1996, Australia was the first country to be 'examined' in the OECD's thematic review on the transition from school to work. Indeed, it lobbied vigorously to be included "because Australia was keen to show what it had achieved in this area" (IN 19: bureaucrat, 1997). At the same time, vocational education has over the years become a priority elsewhere, as being central to developing skills for the global economy. Vocational education and training provides a useful vehicle for thinking about how education policy becomes 'globalised', about the part an international organisation such as the OECD may play in that process, and about the consequences for policy making at the national level.

In 1974 Australia commissioned an OECD study into issues posed by the transition from school to work. The narrative has two points of departure: within the OECD and within Australia. To understand the shifts in the vocational education and training agenda a little backtracking is needed. Here vocational education and training are used somewhat loosely as an umbrella term for a number of related issues: recurrent education; transition from school to work; vocational and technical education; technical and further education; post-compulsory education and training; and most recently, lifelong learning. The terminology reflects both different conventions (e.g. technical and further education in Australia, further education in the UK) and also differences in meaning which changes over time. For example, recurrent education once constituted a strategy *for* lifelong learning, though lately the former has fallen out of use. The notion of recurrent education, however, serves as backdrop to this discussion.

Initial Developments in Recurrent Education Policy within the OECD

Recurrent education first surfaced within the OECD in a publication *Recurrent Education: A Strategy for Lifelong Learning* (CERI/OECD, 1973). The concept of recurrent education was predicated on the belief in distributing education "over the lifespan of the individual in a *recurring way*" (Cantor, 1974, preface, original emphasis). As the citation which headed this chapter makes clear, the concept was radical, even utopian. The outcome of recurrent education, Gass claimed, would be "the individual's liberation from the strict sequence of education–work–leisure–retirement and his freedom to mix and alternate these phases of life within the limit of what is socially possible, to the satisfaction of his own desires and needs" (Cantor, 1974, preface). There were strong parallels to the concept of lifelong education, developed by UNESCO and expressed in the Faure Report, *Learning To Be* (UNESCO, 1972). Both concepts gave voice to "a complete rethinking of education as a system and a process in the context of the other (political, economic, cultural etc.) elements of the total social system" (Duke, 1974, p. 10).

At first, recurrent education embraced the twin objectives of promoting individual development and of providing full-time education for adults as a way to promote inter-generational equality of opportunity (McKenzie, 1983, p. 12), the latter aspect reflecting Scandinavian principles of social equality between generations as well as social groups. CERI spent some time developing the idea. It was, in Papadopoulos's opinion (1994, p. 112), the closest the OECD came to "advocating an explicit strategy of its own for the long-term development of educational systems". However, the concept was attacked. 'Egalitarians' saw recurrent education as a means for overcoming the "dominantly selective function of education systems ... and overcoming socially-determined educational inequalities" (Papadopoulos, 1994, p. 112). Ultimately, the idea sank in face of "fears on the part of the establishment that the application of recurrent education would result in a radical transformation of existing educational systems" (p. 114). As originally envisaged, recurrent education would have challenged the tenets of meritocracy and involved a major revision of administrative arrangements in member countries because of the need for intersectoral policy and administration at the time. The Minister for Education from the UK, Margaret Thatcher, Papadopoulos noted, was a particularly vigorous opponent (p. 114).

The OECD pulled back from its radical stance and retreated to more technical problems of 'bottlenecks' in access to education and training, though almost certainly some ideas relating to recurrent education developed by CERI in the 1970s did subsequently permeate stances on vocational education and training in member countries. Still, the OECD did not take an active position again in this area until the 1990s. How the idea surfaced within Australia around notions of transition education is the subject to which we turn next.

Transition Education Policy within Australia

Australia joined the OECD in 1971, ten years after the establishment of the Organisation and only after the retirement of John McEwan, Deputy Prime Minister and leader of the Country Party, then minority partner in the federal Coalition government. A staunch protectionist, McEwan heartily opposed Australia's entry because of the OECD's free-market stance. Since joining, however, Australia has been a highly active member, hosting more researchers than from any other country (IN 16: secretariat, 1995) and providing the two subsequent Deputy Directors for Education following the retirement of George Papadopoulos — Malcolm Skilbeck and Barry McGaw.

In education, Australia's involvement with the OECD was initially low key. In the 1970s, the Head of the Department of Education, Ken Jones, used the

OECD as a channel to review the dialogue between the Australian States and with the Commonwealth (IN 55: secretariat, 1997). The OECD also served as a policy reference point during the reforming years of the Whitlam Labor government. The influential Karmel Report (Schools Commission, 1973), the Curriculum Development Centre (then headed by Malcolm Skilbeck), and the Schools Commission (an advisory and policy management body to the government) — all drew on concepts of educational disadvantage and social inequality which CERI promulgated during the 1960s and 1970s. A number of Australian educationists went on to chair the examining panels for OECD reviews of national education systems in the early 1980s, for example Ken Jones (Yugoslavia, 1981), Peter Karmel (US, 1981, and New Zealand, 1983) and Greg Hancock of the Schools Commission (Denmark, 1980). In vocational education, the first significant document to propose reforming Australia's Technical and Further Education system, the Kangan Report (Kangan, 1975), drew on CERI's work on recurrent education as well as on UNESCO's *Learning To Be* (Kearns and Hall, 1994, p. 5).

However, according to members prominent in the Schools Commission, the direction of influence at that time flowed from Australia to the OECD rather than vice versa, particularly in the schools sector then taken up by the Schools Commission's innovative Disadvantaged Schools Program (IN 48: bureaucrat, 1996; IT 45: bureaucrat, 1996). The latter, which funded schools focussing on whole school change rather than on individual students, "had a very big effect on the thinking of the OECD ... The Schools Commission was regarded as a place to look to by the OECD" (IT 45: bureaucrat, 1996). Education policy at that time, it has been asserted, was driven by the internal dynamics of Australian politics and by an Australian tradition of reform drawing on ideas "going back into antiquity":

> "schooling was seen as a redemptive exercise, that it could make a difference for kids — that was the driving force. That's not OECD stuff, that's a hundred years of educational philosophical developments ... They go back a long way in the literature."

> (IT 45: bureaucrat, 1996)

Australia did participate in CERI's recurrent education activity. It provided the report by Duke (1974). Ideas on recurrent education developed by CERI did filter into the Kangan Report, though the latter was more explicitly indebted to *Learning To Be*. However, the commissioning by the Whitlam government of an OECD educational policy review on "the transition from school to work and further education" in 1974 marked the beginning of a more active engagement by Australia. The notion of transition education in the 1970s became associated with programs focussing on links between school and work. 'Transed' was often decried in educational circles as narrowly vocational, a band-aid for keeping reluctant school stayers out of trouble during periods of high youth unemployment. When the OECD review on transition education was commissioned, however, youth unemployment was not perceived as a significant issue. The topic was selected because "the Australian education authorities had been conscious not only of their own shortcomings in this area but of the increasing attention that their OECD partners were paying to this subject" (OECD, 1977, p. 85). Initial interest in the topic was then broadly framed. Transition to work was regarded as an important area in view of the importance work had in allocating adult status and identity. In this, links with prior work on recurrent education played their part.

As suggested in Chapter 3, OECD reviews of national education systems have been a significant part of the Organisation's work. The review process usually involves preparing a background report by the country under examination, a visit by the review team, an 'examination' or 'confrontation meeting' in Paris between country representatives and the OECD review team. Finally the report is published. In this specific instance, a working party was set up to provide information for the background report. It adopted the broad framing noted above: "The vocational development of an individual has to be seen as part of his total human development ... the emphasis should be on orientation to the world of work or vocational preparation generally, rather than training for a particular job" (Commonwealth Department of Education, 1976, pp. 91, 118).

The deliberations of the working party were cut short by a change of government in 1975. Though the background report, finally produced to inform the OECD review team, demonstrated the beginnings of a broader approach, by the time it appeared youth unemployment had become a central preoccupation, distorting to some extent the original emphasis (OECD, 1976). The OECD review itself (OECD, 1977, p. 16) commented on the changing context. It noted the "increasing unease" about the social role of education compared with the "heady optimism" of the 1960s. Consequently, the review interpreted transition in "both a broad and narrow sense". It focussed on specific issues facing young people, and particularly early school leavers, "at the interfaces between the systems of education and employment". More broadly, it proposed to study the linkages between education and employment systems, on the grounds that:

> "It is obvious that coherent policies to facilitate the transition from school to work call for an adequate degree of coordination between education and employment policies. The examiners came to the conclusion that, as in many other countries, such coordination needs to be reinforced. That the Australian authorities share this view is shown by the fact that a national 'Committee of Inquiry into Education and Training' has now been appointed."
>
> (OECD, 1977, pp. 5–6)

Before the OECD review was finalised, the new Fraser Coalition government set up a Committee of Inquiry into Education and Training (the Williams Inquiry) to consider the relationship between education and employment. However, the Williams Report *Education, Training and Employment* (Williams, 1979) made only a brief reference to the OECD review's recommendations for better coordination. It was silent on matters pertaining to educational structures or to policy integration (for example, of education and labour market policies). By and large it adopted a relatively narrow approach to the question of transition. Although the Williams Report was careful not to "blame" schooling for youth unemployment, suggestions that schools could contribute to the skills and attitudes of young people "by the encouragement of a disciplined habit of work" (p. 133) made it politically controversial. The peak teachers' union, amongst others, protested against the Report's "deficit" stance (Roche and Marginson, 1979). In general, it was seen by educational circles as highly conservative and likely to preserve existing divisions and hierarchies. However, it did resonate with the times. It provided the impetus for the establishment of a Commonwealth Transition Education Program which funded initiatives in work experience and the provision of 'alternative' vocationally oriented subjects in the largely academically oriented secondary schools.

Changing Contexts and Policy Approaches

By 1977 youth unemployment had become a major concern and, influenced by work in the UK by the Manpower Services Commission, the Australian Education Council (a body comprising the commonwealth and state ministers of education) became interested in training as a means of responding to the long-term, structural nature of youth unemployment, a notion increasingly taken up in the 1980s by all OECD countries. The next phase of thinking about transition issues, both in Australia and the OECD, focussed on broad issues related to youth unemployment and policy against a background of labour market restructuring. As Margaret Vickers' discussion of OECD's influence on Australian education policy showed (Vickers, 1994), the new approach was evident in the various reports that accompanied Australia's second OECD country review on youth policy initiated in 1983 by Peter Wilenski, Secretary to the Department of Youth Affairs in the new Hawke Labor government.

The previous year, Wilenski (a former Department Head in the Whitlam Labor government) had been invited to work at the OECD in Paris to develop a comprehensive paper on youth policy. Amongst other things, his paper recommended an integrated approach to youth policy covering education, training, employment and income support — ideas which were reflected in Australia's background report for the OECD Review (Department of Education and Youth Affairs, 1983, p. xxxii). The background document referred to problems with the transition education approach contained in the Williams Report. It asserted that the goal of increasing participation and equity was "a key element in the new framework of youth policies" (p. 25). The OECD review (OECD, 1986b) was consistent with this broad-based approach. It recommended a "youth entitlement" and the development of comprehensive policies for youth, with emphasis on the long-term needs of young people rather than on short term concerns of high unemployment. After 1983, Wilenski began to implement new policy directions in keeping with those recommended in his OECD paper. He set up an Office of Youth Affairs, established a review of income support arrangements for young people which led to the AUSTUDY (a student support) scheme and helped to initiate the important Kirby Review of the Labour Market which recommended mainstream education and training to replace earlier labour market programs. In her account Vickers (1994) suggests that the OECD helped youth policies to be framed in a new way. She argued that Wilenski's experiences at the OECD and " … the combined weight of the OECD review together with the confluence of views expressed in the Kirby report and the income support review created a climate in which the government's plans could be enacted" (p. 41).

At that period there was something of an ideological struggle within the Labor government (preceding its 'new Labour' British counterpart by a decade), a battle between old-style 'wets', associated with the former Whitlam era, and economic 'dries'. As Education Minister until 1987, Susan Ryan, a prominent feminist and a wet, did much to promote a strong social orientation for education, reinvigorating the Schools Commission, and initiating important equity programs such as the Participation and Equity Program which replaced the former Transition Education Program. In 1987, with the ascendance of the economic dries, traditionally separate policy areas were combined in a mega-Department of Employment, Education and Training (DEET) with John Dawkins as Minister. The amalgamation sought "to achieve a new coherence and consistency between our various education and labour market policies and programs" and formed part of what Dawkins described as a "vigorous programme of micro-economic reform" in Australia (OECD, 1989b, p. 10).

Such directions had been vigorously promoted in various OECD reports (e.g. OECD, 1985c, 1987e, 1989b). However, as Dawkins himself commented, the creation of a single portfolio was a first among OECD countries. The formation of DEET was seen as "a milestone in relations with the OECD" (IN 65: bureaucrat, 1996). The new DEET structures "led to a greater policy focus and interest in the OECD" (IN 44: bureaucrat, 1996), and to "a tighter control over education agendas for the OECD … compared to the old Education Department" (IN 31: consultant, 1996). In the words of one consultant (IN 35: 1996): "After Dawkins, the OECD links were cranked up — DEET was more active in seeking information. The riding instructions from Dawkins were stronger."

Thus from the late eighties onwards, the OECD seems to have played a more visible role in Australian policy making. This was particularly evident in relation to the government's agenda for restructuring. Vickers (1994), for example, argues that during this period Canberra's education bureaucrats focussed selectively on those aspects of the OECD agenda which emphasised the economic functions of education. Looking back on that time one consultant observed (IN 43, 1996): "Their [DEET's] key players had graced the OECD and saw in the OECD a conceptual, ideological framework of organisation which could legitimate its work." Dawkins himself used the forums of the OECD to propound his own version of economic rationalism, namely the pivotal role of education and training in skills development for the global economy, the need for educational restructuring to achieve this end, and the complementarity of economic and social purposes of education. According to Vickers he also drew on OECD reports to legitimate his stance. As an illustration she highlighted Dawkins' use of *Universities Under Scrutiny* (OECD, 1987d) which hewed out the path towards closer relationships between governments, institutions and firms. Dawkins cited this report twice in his foreword to the Green Paper on restructuring higher education, but his line of argument, as she suggests, interpreted the statements in the document "not as a simple description of trends, but as a rallying call to action" (p. 42).

In 1988, Dawkins chaired the OECD conference 'Education and the Economy in a Changing Society', a conference which took as its theme the convergence of education and economic functions in the new global context, a theme which rested on what was becoming an increasingly familiar argument of workers' skills and qualifications "as critical determinants of effective performance of enterprises and economies" (OECD, 1989b, p. 18). While these were also Dawkins' own ideas, he was able on his return from Paris to use the conference to legitimate and promote his educational restructuring agenda more strongly in Australia (Vickers, 1994, p. 43).

Australia and the OECD: Merging Agendas

Inspired by the 1988 conference, the OECD initiated a three year activity, 'The Changing Role of Vocational Education and Training' (VOTEC). It aimed at examining approaches and programs in school-linked vocational education and included links to workplace training and relationships between general and vocational education. This activity represented the second period of active commitment by the OECD to issues of vocational education (the first having been to recurrent education two decades earlier). It was a significant departure from its chief preoccupation with higher education (IN 50: secretariat, 1995). Australia, having made vocational education and training a policy priority under the agenda of national training reform, was an active participant in the VOTEC project.

Both in the OECD and Australia, the equity-enhancing aspects of vocational education and training were stressed. In both, the concern was to avoid a narrow interpretation of vocationalism. In Australia, reforms to vocational education and training in the early and mid-1990s laid weight on bringing together general and vocational education so that general education was seen as relevant to work and vocational education was perceived as broader than specific work-based skills. The reforms also underlined the importance of creating flexible pathways between school, TAFE (Technical and Advanced Further Education), higher education and work to include virtually all young people. Although the reforms were criticised for their symbolic and simplistic attention to disadvantaged groups (Taylor and Henry, 1994), the essential argument remained that the various policies initiated under the training reform agenda would lead to more generally equitable outcomes than had previously been the case precisely because of their more inclusive education and training provisions.

Similar ideas underpinned the OECD's VOTEC activity. So, for example, its final publication noted "there should not be complete separation between general education and vocational training, and that as far as possible general education should continue during vocational training" (Pair, 1998, p. 23). Also noted was the underlying assumption that:

> "in a democratic society, policy makers have an interest in participation in VOTEC ... there may be economic reasons for this policy interest ... There may also be social reasons: for example, the perception that participation in VOTEC may prevent failure and exclusion, especially among disadvantaged groups."
>
> (Raffe, 1998b, pp. 376–377)

How far equity objectives and broadly defined vocational education and training goals were realisable — or realised — is another matter. In Australia, scant attention was paid to analysing and addressing the sociological, psychological and historical factors underpinning the divide between general and vocational education, hence the attempt to bring these areas together somewhat appeared more rhetorical than carefully constructed. Equity objectives, while strongly expressed, were in fact strongly resisted and poorly implemented with a lack of vigilance in the monitoring processes and in the establishment of a more deregulatory training climate which helped to reduce monitoring capacity further (Henry and Taylor, 1995; Taylor and Henry, 1996). Within the OECD, sociological and psychological as well as economic factors having an impact on vocational education and training reform were taken into account in many of the papers contributing to the VOTEC-related conferences and helped to ground its core concept of pathways as a way of thinking about bringing together general and vocational education and to do so more inclusively (OECD, 1994g, 1998e). In contrast, too, to directions pursued in Australia, the Organisation argued that "the desired pattern of participation will not necessarily be achieved by ... leaving everything to market forces. A *laissez-faire* approach is unlikely to address issues of equity, the needs of the disadvantaged or the long-term interests of the economy" (Raffe, 1998b, p. 377).

In 1995, building on the VOTEC activity, the OECD launched another project, 'Transition from School to Work: the roles of general and vocational education', renamed in 1997 as 'Improving School-to-Work Transition' as part of the mandate of 'Lifelong Learning for All' (OECD, 1996d). The thematic review of the transition from initial education to working life was part of this project. Australia was the first country to be reviewed, thereby completing the circle begun in 1976 when it "pointed the way" by making the transition from school to work the focus of its first OECD review.

But a circle is perhaps not the appropriate metaphor to describe the policy developments which have been analysed here. Rather, what emerges is a chain of policy development, a mix of national, international and global elements and an interplay of political, ideological, economic and labour market factors which impinge both on national policy making and on the work of the OECD itself. This *mélange* is highlighted in the Australian experience as it has been recounted. Here, three phases of policy development can be seen: the early phase of transition education, coinciding with the first OECD review and with a shift from a Labor to a Coalition government (1974–1983); the middle phase in which more coordinated approaches to youth policy were developed (1983–1987), coinciding with the second OECD review and a Labor government in transition; and the third phase (1987–present) of educational restructuring, coinciding with the OECD thematic review on transition from school to work, and a shift from a Labor to a Coalition government.

In the first phase, the OECD appeared a less significant source of policy ideas though the then Labor government clearly regarded the Organisation as sufficiently useful to commission an education policy review. Arguably the Schools Commission's work on educational disadvantage was the most significant and innovative policy legacy of that time. While that undertaking acknowledged ideas drawn from OECD reports, it drew more substantively on a long 'Australian' tradition of thinking and from the experiences of the US and the UK. In terms of the vocational education and training agenda, the OECD review may have assisted Australian authorities "by articulating problems and developing something of a national perspective". It may have had a catalytic effect in the sense of "establishing a basis for and encouraging further investigations and discussion" (Kogan, 1979, p. 66). But its policy prescriptions were broadly couched, enabling differing interpretations of transition education. Given changing economic and political circumstances, a narrower version than may have been initially intended was taken up. Indeed, it would seem that in this phase the Williams Report — a creature of the then conservative Coalition government — had greater weight for educational policy directions than did the OECD review.

During the middle phase as we saw, links between Australia and the OECD were more closely shackled together. The structural nature of youth unemployment was beginning to be better understood. This period marked the shift to an economically 'drier' climate, although education under Minister Susan Ryan retained a predominantly social orientation until 1987. During this period, much work was done to try to implement Wilenski's work on youth policies. For this the OECD was used for purposes of legitimation and enlightenment (Vickers, 1994). After 1987 and the formation of the Department of Employment, Education and Training, the final phase in this chronology, policy priorities pointed more directly towards the imperatives of a globalising economy. This period also marked a high point in Australia's involvement in educational matters within the Organisation — a far cry from McEwan's protectionist opposition to the OECD. Dawkins, as Education Minister and then Treasurer, used the forums of the OECD to promote and legitimate his policy prescriptions for educational restructuring and microeconomic reform — prescriptions which fitted the OECD liberal economic template like a glove, particularly the emphasis on human capital formation. Indeed, Dawkins helped to shape the model. Under this 'drier' Labor regime, notions of equity were harnessed to economic rationalist goals of efficiency. They reflected a similar grafting of liberal democratic ideas onto market liberal principles in the OECD, noted in Chapter 4. Replicating the fate of earlier initiatives in transition education, the policy agenda of training and skills formation survived a change of government and, in a replay of earlier events, although it was Labor which lobbied to participate in the OECD's thematic review on transition education, the review itself was conducted under the mandate of the Coalition government in 1996.

By the 1990s, however, party political commonalities rather than differences around the "new policy paradigm" of human resource development and market liberalism (Neave, 1991) were increasingly evident. Both sides of politics agreed on the necessity of a reduced role for the state in education, of a market-driven system of provision, and what Bienefeld (1996, p. 429) referred to as the "cargo cult" of vocational education and training as the solution to the volatile demands of global labour markets. Relatively minor differences existed over how vocational education and training should be funded and how questions of access and equity should be conceptualised (Taylor and Henry, 1996). In bending to this general trend, Australia has been but one amongst many countries which followed a similar policy path, albeit each with its own distinctive characteristics (Green, 1999).

Where does such a path begin? It is difficult to argue that it springs entirely from an Australian tradition of reform, though there are undoubtedly specific indigenous elements in the way the ingredients are mixed. The agenda for national training reform drew upon on a series of local documents (for example, Australian Education Council Review Committee, 1991; Australian Education Council/Ministers of Vocational Education, Employment and Training, 1992; Employment and Skills Formation Council, 1992) which collectively framed a distinctive Australian policy framework for vocational education and training. Yet, there are elements of policy adaptation. The training reform agenda had its genesis in a mix of Swedish and UK models of vocational education and training.

However, more is at stake here than cross-national comparisons and adaptations. Rather, we would argue that a convergence of policy ideas is taking shape, a convergence emanating from the machinations of increasingly interlinked policy networks and supranational and international organisations — an emergent global policy community — coalescing around key agendas. Vocational education and training, because of its explicit links with the economy, is one of these agendas. An early example of creating such a supranational arena is provided by Neave's (1991) discussion of the wedge into national policy-making created by the *entrée* of the European Commission (EC) into higher education policy in the mid-1980s via the Trojan horse of vocational education and training. Motive for the Commission's involvement was the concern that a shortage of skilled labour might result in European economies falling behind those of the US and Japan. The means by which the EC could intercede in higher education policy was provided by a judgment of the European Court of Justice in 1985, the upshot of which was that higher education be considered a sub-set of vocational education. The judgment brought higher education firmly within the ambit of the Treaty of Rome under which the European Commission operated. As a result, Neave claimed, a new normative level of policy making was established, with the EC being instrumental in reframing student mobility programs away from a concern with cultural exchange to a more utilitarian interest in the "mobility of high-level manpower between industry and across frontiers" (p. 38). One could broaden Neave's conclusion further. Higher education became an explicit policy instrument for European integration. One might equally argue that vocational education and training for human resource development has become an explicit policy lever for global economic and labour market integration.

From this it would follow that while national policies at one level reflect particular national or sub-national histories and priorities, at another level policies are framed by agendas increasingly convergent around human resource development advocated by such influential international organisations and agencies as the EU, UNESCO, the World Bank and the OECD:

> "The level of the competence of a country's skilled workers and technicians is centrally important to the flexibility and productivity of its labor force."
>
> (World Bank, 1991, p. 7)

> " ... higher education is moving towards a mass enrolment system as modern economies become increasingly knowledge-intensive and therefore depend more on graduates of higher education, who constitute a 'thinking work-force'."
>
> (UNESCO, 1994, p. 24)

> " ... learning, expertise, and human resources — the 'human factor' as it is sometimes called — are critical elements of the well-being of our economies and society."
>
> (OECD, 1992e, p. 14)

The significance of convergent rhetoric should not be overstated given the very different ideological stances, purposes and policy settings of these international organisations. Nor should we lose sight of the fact that international organisations like the OECD ultimately reflect the policy priorities of their constituents. Nevertheless, there is a sense in which external context concentrates the policy agenda as we noted in Chapter 3. As a result:

> " ... you'll find the approach is very similar, not because there's a deliberate attempt to do that but because people working in the same field with the same knowledge basis, as it were, the same linkages, the same connections, and often consultants from countries will work for more than one of those organisations. So there's an increasing commonality of those policy interests ... "
>
> (IT 56: secretariat, 1995)

In the words of one OECD delegate, "the OECD and the EC have almost identical agendas ...   I've just come back from an EC meeting at Brussels. The same issues come up" (IN 60: bureaucrat, 1995). One such issue is lifelong learning, the most recent episode in the vocational education and training saga.

Lifelong Learning for All: Converging Policy

Today, lifelong learning or lifelong education has acquired wide currency. It represents a policy priority across many countries and is one of the mobilising slogans for UNESCO, the European Union and the OECD alike. In 1994, Lifelong Education for All became a major term of reference for the UNESCO Mid-term Strategy over the period 1996–2001. In 1996, the OECD Ministerial conference on education adopted 'Making Lifelong Learning a Reality for All' as the theme of its mandate for 1997–2001. In 1995 the European Parliament declared 1996 the 'European Year for Lifelong Learning'. Not surprisingly, the starting point and emphases of these Organisations are somewhat different given their different charters. So, the setting out of lifelong education in the UNESCO-commissioned Delors Report (Delors, 1996) in terms of "the necessary utopia" is closer philosophically to UNESCO's 1970s conceptualisation of lifelong learning described at the beginning of this chapter, than it is to the OECD's more pragmatic and current concerns. As one member of the OECD secretariat commented:

"UNESCO's talk is more about core humanistic values ...   it's more utopian ... It's not that we don't have our dreams, it's simply that a document like that would *never* go down in the Education Committee. Our paymaster simply expects something different. We're not in the same street as UNESCO."

(IT 17: secretariat, 1997)

For all that, there is considerable mismatch between the philosophical starting point of the Delors Report and the position eventually espoused by UNESCO. In fact, OECD and UNESCO rhetoric is remarkably similar. Both organisations endorse uncontestable propositions about the contribution of education to personal, social and economic development, about a "learning society", and about flexible pathways encouraging equitable access and participation in education, now held to extend beyond formal institutions (OECD, 1996d, pp. 15–19; UNESCO, 1996, pp. 17–19).

Underpinning these positions is the faith in education as a means — perhaps the prime means — of not only providing the changing skills required for an information-based economy but, more broadly, of promoting social cohesion and personal development. Accordingly, the stakes are high:

"A new focus for education and training policies is needed now, to develop capacities to realise the potential of the 'global information economy' and to contribute to employment, culture, democracy and, above all, social cohesion. Such policies will need to support the transition to 'learning societies' in which equal opportunities are available to all, access is open, and all individuals are encouraged and motivated to learn, in formal education as well as throughout life."

(OECD, 1996d, p. 15)

"There is a need to rethink and broaden the notion of lifelong education. Not only must it adapt to changes in the nature of work, but it must also constitute a continuous process of forming whole human beings — their knowledge and aptitudes, as well as the critical faculty and the ability to act."

(UNESCO, 1996, p. 19)

Lifelong learning, envisaged thus, is held to advance through a system-wide network of 'learning pathways', extending from early childhood through to all stages of adulthood in both formal and informal educational settings, fulfilling "social and economic objectives simultaneously by providing long-term benefits for the individual, the enterprise, the economy and the society more generally" (OECD, 1996d, p. 15).

Within the OECD, the agenda is scarcely a new one, evolving out of earlier approaches to vocational education, training and recurrent education. The discussion thus returns us to the starting point of recurrent education, initially seen as a strategy for lifelong learning. With the collapse of the egalitarian ethic, however, and the change in economic circumstances, the concept was considerably reworked by CERI. "By the beginning of the 1980s ... to [recurrent education's] long-standing missions of emancipation and equality of opportunity, [has] now been added development of human resources for economic recovery" (OECD, 1987c, p. 4). In 1992, lifelong learning was included as one of the Directorate's medium term priorities (OECD, 1991) followed in 1993 — the year that 'training' was added to the title of the education division — with a new CERI activity on lifelong learning. The objectives of that activity "to match supply of education and training opportunities with the demand patterns identified in work to date" (CERI, 1994f, p. 11) show how far the agenda of recurrent education had moved since its early idealistic expression.

However, OECD educational agendas rarely reflect economic imperatives in a straightforward manner, and this is true of lifelong learning. As one member of the CERI secretariat explained:

"Today, there is a new opportunity to revise the balance between initial and lifelong education and training. A new political climate has been created, first and foremost for reasons connected with economics ... It is unlikely that governments would today give a new priority to education and training for adults were it not for this economic perception. Yet it would be a great mistake to see lifelong learning too narrowly in terms of the skilling and re-skilling of workers. One should not lose sight of the humanistic aims of education, nor regard them as competing alternatives to utilitarian aims. As concluded by the chair of OECD's 1991 intergovernmental conference on further education and training of the labour force: 'the different orientations are not mutually exclusive, and are in fact mutually reinforcing'."

(OECD/CERI in Atchoarena, 1995, pp. 201–202)

Interesting in this context is how far member countries such as Australia, have come in terms with "a radical transformation of existing educational systems" (Papadopoulos, 1994, p. 114), a condition seen as a prerequisite for implementing the original idea of recurrent education. Such a transformation may not have overcome the "dominantly selective function of education systems" (p. 112); but the previously insurmountable obstacle of intersectoral policy development and administration is, if not resolved, at least generally well recognised and accepted. Some of the rethinking involved may have been a result of the OECD's catalytic role and the consistency with which it argued the need for better cooperation and coordination between education, labour and social administrations. At the same time, much of the transformation may have come about less for the reasons advanced by the egalitarians, but rather as a result of the cross-bonding of education with other policy fields as the pressures of global competition and human resource development started to weigh upon educational priorities. An illustration of this process may be seen in an Australian report, *New Media and Borderless Education* (Cunningham et al., 1997), commissioned by the Department of Employment, Education, Training and Youth Affairs. It highlighted the place of educational exports for the Australian economy, and acknowledged "the increasing importance of cross-sectoral collaboration, alliance and partnerships". In light of these factors it recommended that "Australian governments should develop cross-portfolio policies and strategies addressing the inter-relationship of higher education, telecommunications, information technology and mass media policies . . . " (unpaged). Economics, it would seem, could mobilise systems in ways that ideals could not.

While continuities exist between the earlier, radical agenda and current initiatives in lifelong learning, discontinuities are notable. Partly because it emphasised formal education, the strategy of recurrent education assigned a large role for government. Nowadays, continuing work-based vocational training is regarded as preferable to formal adult, institution-based education. With this change in perspective has come a concomitantly lesser role for government in organising, managing and financing the system and an "increased reliance on the responsibilities of employers and individual learners" (OECD, 1996d, p. 89). On the conceptual level, too, the notion of 'social demand', implicit in the earlier version of recurrent education, has been replaced by 'individual demand' as a key concept. In other words, lifelong learning is now interpreted not "as a right to be exercised, but as a necessary requirement of participation" in economic life and hence civil society (p. 89).

These shifts reflect the contradictory philosophical foundations of contemporary education (and particularly post-compulsory education) systems. The dominant 'selection function of education' remains firmly entrenched in the new competitive order, and pertains as much to the performance of individual students as to institutions, as we saw in Chapter 5. But it has been overlain with the ostensibly more egalitarian objective of participation for all. This latter has been bolstered by recognising the socially divisive effects of economic globalisation: "Beyond the search for increased competitiveness, investing in lifelong learning is also viewed as a way of maintaining or rehabilitating social cohesion" (Atchoarena, 1995, p. 217). Whilst such objectives still resonate with the earlier ideals of recurrent education, the realisation of such objectives remains, we would argue, problematic given the discordant logics jangling uncomfortably against each other inside the discursive framework of lifelong learning. The contending logics are the competitive impetus of efficiency and performativity against the egalitarian impetus of participation for all. The Delors Report (1996, p. 15) identified this tension in the form of "the need for competition" and "the concern for equality of opportunity", a tension for which no solution has "stood the test of time". That Report, too, laid out in stark terms the social devastation accompanying the march of globalisation:

"It may therefore be said that, in economic and social terms, progress has brought with it disillusionment. This is evident in rising unemployment and in the exclusion of growing numbers of people in the rich countries. It is underscored by the continuing inequalities in development throughout the world ...   The truth is that all-out economic growth can no longer be viewed as the ideal way of reconciling material progress with equity [and] respect for the human condition."

(Delors, 1996, p. 13)

While such a backdrop is commonly acknowledged, it is rarely laid unambiguously upon the globalisation ideal of economic growth. On the contrary, there is an apparent unwillingness to confront the inherent tensions within the rhetoric of lifelong learning. On the one hand, lifelong learning may be reduced to being a potential source of income:

"There is a widespread perception that traditional institutions are not meeting the needs of the lifelong learning cohort and that the field is open for new providers to meet market demands. One obvious, and problematic, outcome of this segmentation is that traditional institutions may be left serving the less-profitable traditional undergraduate market which is largely government-funded or family-funded, in a time when governments are increasingly endeavouring to cut public outlays ... [Therefore it is recommended that] Australia's higher education sector ...   develop strategies to compete successfully in the profitable lifelong learning market."

(Cunningham et al., 1997, unpaged)

On the other hand, lifelong learning may be reduced to the level of cliché:

"Ministers are keenly aware that lifelong learning for all will require additional financial resources. But because lifelong learning provides substantial economic and social returns to all partners — individuals, families, employers and the society as a whole — the additional investment must be mobilised by all concerned."

(OECD, 1996d, p. 14)

"Education can no longer be conceived of as a one-chance affair ...   It should be seen as a continuing process whereby individuals are offered learning opportunities not just once but a number of times throughout their lives ...   Advancing towards lifelong education for all implies moving towards a 'learning society' in which each person is a 'learner' and at the same time a 'source of learning' ... "

(UNESCO, 1996, p. 18).

Whether prescriptions for lifelong learning are more likely to resolve the tensions between equity and efficiency identified in the Delors Report than previous 'solutions' is perhaps an open question:

" ...   there is a lot of pressure from international development, from the private sector, from the IT industry which certainly is going to have a major impact on how lifelong learning is going to take shape. The big issue is, is it going to be lifelong learning for a few or ... and there the jury is still out ...   But there's a big chance, a big chance."

(IT 49: secretariat, 1997)

The interest in lifelong learning constitutes only a segment of the broader agenda of vocational education and training. The concluding section of the chapter will return to this broader theme the better to consider the question with which we started, namely the implications of the globalising drive in education policy for the meaning of 'national' policy making.

Implications for National Policy Making in Education

Prunty's definition of policy as the "authoritative allocation of values" (Prunty, 1984) provides a useful starting point for thinking about this question. At issue are two sub-questions: "where does this authority come from?" and "whose/which values are being allocated?" In the past, the most obvious answer was that authority derived from the nation-state or a sub-national political unit within the nation-state since it is at these levels that educational policy is made and funded for implementation. The values enshrined in policy are those reflecting the dominant discourses — and the political compromises — within nation-states at any moment. This chapter has developed one of this study's central argument. This is that the nation-state's authority in the allocation of values has not ceased. Rather it increasingly sits alongside other value-allocating authorities. In the words of Hirst and Thompson (1996, p. 183): "politics is becoming more polycentric, with states as merely one level in a complex system of overlapping and often competing agencies of governance". With the momentum behind globalisation, both materially and ideologically, making policy "in the national interest" has become, analytically and politically, a complex issue.

Theorists of globalisation argue that globalisation exerts simultaneous expressions of convergence and fragmentation, of universalism and localism. These tendencies were seen in the Australian case study presented here and most especially in the way nationally developed policy stances centred on vocational education and training existed alongside, and become enfolded into, the new global policy paradigm of human resource development. Commenting on Australia in relation to the thematic review on the transition from education to work, the OECD observed:

> "Australia is not alone in experiencing youth unemployment, social alienation among the young, and concerns about longer-term economic prosperity. However, in some respects the forms that those concerns take are unique to Australia, as are some of the policies that have been launched in response. As a society that is becoming increasingly oriented to the Asia–Pacific region, and which has strong ties to Europe and North America, the Australian experience has much to offer the OECD membership as whole."
>
> (OECD, 1997e, p. 1)

At one level Australia, like all countries, upholds its own distinctive policy stance on vocational education and training, reflecting a culturally and historically specific mix of factors. At another level, Australia contributed to, and its policies reflect, what may be called a globalised policy discourse of vocational education and training in which national differences are muted, with the OECD acting in this process as an "international mediator of knowledge", as described in Chapter 3. The expression of strongly articulated national sentiments may indeed be one aspect of, or response to, globalisation. Here it is worth noting that the policy paradigm of human resource development is grounded in notions of *national* competitiveness in global markets. On the basis of observation, one might revert to the pluralist conclusion, implicit in the OECD's *modus operandi*, that countries do possess a strategic space for autonomous policy making. While at one level that is true, pluralist assumptions ignore the power of dominant discourses to incorporate and relativise challenges. Thus, the mantras of market liberalism and human capital investment remain virtually uncontested within both global and local policy communities, although there is a body of critical research and literature which most certainly casts doubt on some of the assumptions underpinning that policy paradigm. Marginson (1997b, pp. 117–118), citing a number of studies, referred to the "popular narrative of investment in education" as one of the "great modern myths, transcending the need for empirical verification". In a similar vein, Ashton and Green (1996, p. 3) argued that "it is incorrect to assume a linear and automatic connection between skill formation and economic performance", that links between training, profitability or economic growth "are still largely in the realm of theoretical belief or just plain hope" and moreover that the notion of a globally integrated economy is simplistic (p. 5). Such views, however, remain very much in the economic and political shadows.

Yet, it is possible to argue in favour of a strategic space for policy making "in the national interest" and to do so in ways which both recognise the complexities involved in conceptualising the national/global interface and guard against what Hirst and Thompson (1996, p. 2) qualify as "the pathology of diminished expectations" of the state. Marginson (1999) reminds us that the nation-state still retains an impressive portfolio of policy responsibilities. It is still, in other words, a significant allocator of values. Educational policy priorities are thus as much politically derived as ideologically determined. In the case of Australia, as elsewhere, there is a good deal of ideological convergence around vocational education and training. But this study has also shown that there is at the same time *national* politics and sets of priorities, even in a federated state like Australia. To some extent politics and priorities have been sharpened by the cross-bonding of economic and education policy, which is itself symptomatic of the pressures from globalisation. Such a configuration has implications for the OECD's mode of operating, building up tensions between its function as a forum, as a think-tank and its increasingly more prescriptive function as an actor in globalisation.

7

Redefining University Education

"OECD has traced and analysed the moving scene by means of conceptual models and carefully designed working frameworks, by classifying and categorising phenomena, by synthesising research findings and by coining descriptors to facilitate intergovernmental discourse and permit comparisons across countries."

(OECD, 1993c, p. 12)

"The concern with higher education has undoubtedly been vindicated by the impact of the numerous activities on national policy-making, on institutions and on academics researching into higher education as a discrete field of inquiry. In the international context, OECD has long been viewed as a fount of original ideas and helpful prescriptions, the chronicler of the rapidly changing priorities and problems of higher education considered as a total sector, and as an indispensable source of reliable comparative data and up-to-date information derived from direct country contributions and constant monitoring of the relevant literature."

(OECD, 1993c, p. 12)

This chapter looks at the OECD's changing conceptualisation of higher education. It pays particular attention to the ways in which the Organisation has constructed the policy context of globalisation for higher education. Seddon (1994) has argued that policy often relates to creating a context for policy. Policy making is as much about problem creation and setting as it is about solutions to problems (Taylor et al., 1997; Yeatman, 1990). Policies construct their own context. They frame policy problems in ways that suggest the policy put forward will be able to solve the problem as it is constructed and stated.

The perspective adopted here interprets policy in terms of discourse, that is — a struggle between meanings and over meaning (Yeatman, 1990). Policy always involves suturing together competing discourses which represent competing interests. Policies as texts are in turn constrained and framed by the broader discourses within which they are located (Ball, 1994). Chapter 4, which outlined the ideological tensions within the Organisation, and the case studies presented in Chapters 5 and 6, all show that this holds good for OECD Reports as it does for educational policies at the national level.

A useful approach to policy analysis is to consider context, texts and consequences and also to recognise that policy involves both processes and text (Taylor et al., 1997). Policy embraces all of the political and bureaucratic exchanges, interactions and machinations that issue setting engenders, in text production and in the implementation or practice of policy. Ball (1994, 1997), and with his colleagues Bowe and Gold in Bowe et al. (1992), argued in favour of an interactive policy cycle approach to policy analysis. This approach rejects a linear top-down account of policy processes. Instead, it proposes three contexts in a policy cycle: a context of text production, a context of influence and a context of practice. Each sits in interactive relationships with the other contexts. In the case of state-derived policies, more emphasis may have to be laid upon the state as the locus of policy text production. As earlier arguments suggested (see Chapter 2), the rise of globalisation and its various manifestations supports the notion that educational policy analysis now needs to take global and supra-national influences on each of the three contexts into account. The OECD is one important source of such influences.

At one level, our interest is about how the OECD exerts influence in higher education. According to the citation at the head of this chapter, one way the OECD brings its influence to bear is through discursive interventions. The work of the Organisation contributes to the discourses which frame higher education policy in any particular member nation. It creates 'conceptual models', classifies and categorises phenomena, coins descriptors in the policy field of higher education, all intended to 'facilitate intergovernmental discourse'. These discursive interventions are set against its prognostic documentation of trends in higher education. In turn, this documentation works as an agent of "anticipatory policy convergence" (Dale, 1999, p. 13). As the second citation at the head of the chapter made plain, the OECD also provides prescriptions for policy. By focussing on two aspects of the Organisation's higher education work these processes may be made explicit. The first is the Thematic Review — *Redefining Tertiary Education* (1998) — the first of its type conducted by the OECD. The line of argument developed here will be that *Redefining Tertiary Education* is now part of the context of influence in national production of education at tertiary level and that it stands simultaneously as part of the context of influence in policy development at the level of the individual institution.

A number of reasons prompt the choice of the Thematic Review as the focus of this analysis. It was the first such review conducted and for this reason stands apart from the usual approach of nation-based reviews. Indeed, the use of Thematic Reviews was one strategy the Organisation used to attract member countries back into the review process (IT 56: secretariat, 1995) despite some initial resistance from the Education Committee. Interestingly, in a 1979 review of the Country Reports for the OECD, Kogan warned against the temptation of "facile comparison" in thematic, cross-national reviews and argued against blending "national histories and educational policies" into "generalised metahistory" (p. 75). By the early nineties, global geo-politics had so radically changed that Thematic Reviews appeared to be the way ahead for OECD educational reviews. Whilst universities to some extent have always been part of an international community of scholars, globalisation has opened the door to the view of university education as a tradeable item on global markets and has been reflected in the greater flows of students across frontiers.

The Thematic Review provides one focus to this chapter because it goes some way to demonstrating a central argument in this study, namely that the OECD, while being simultaneously a policy instrument, a forum and an actor, has increasingly assumed the role of policy actor. The Thematic Review admits this interpretation when it suggested that it combined an admixture of the "descriptive, analytic and normative" (OECD, 1998c, p. 18). This development significantly moves the OECD beyond a "catalytic and integrative" function vis à vis policy production in its member countries.

The second prism through which one may seek to cast light on the way the OECD brings its influence to bear in higher education is to be seen in the internationalisation of higher education. For most of the 1990s, the OECD actively promoted an agenda of internationalisation of higher education, within its member countries and beyond, notably in Asia. Its interventions in the internationalisation policies of member countries have been discursive, contributing to the framing of their debates on policies. These have enabled countries to build upon their own histories, and, at the same time, to assimilate a new discourse between governments that reconciles competing policy agendas. In Europe, policies of internationalisation built around the success of student exchange programs such as Erasmus and Socrates, which stressed the importance of student and staff mobility and of internationalising the curriculum. In contrast, for countries like Australia, the UK and Canada, policies of internationalisation emerged from a range of commercial concerns, designed to secure a deteriorating financial base. Faced with these competing discourses, the OECD's distinctive contribution involved to moving its member countries towards a blended policy language of internationalisation that reconciled educational and commercial imperatives. Through its attempts to facilitate dialogue, it took on the role of a policy actor, sponsoring an overtly normative agenda, the purpose of which was to urge universities towards a more global outlook and to attend to preparing their students for work in a global economy.

The OECD's Interest in Higher Education: A Brief Chronology

In a self-description of its involvement in higher education, the OECD (1993c) suggested that no other educational sector received so much of its attention — and that from the Organisation's very start. From one standpoint, this attention reflects the coincidence of the OECD's creation with the turmoil in the universities during the sixties. Such a coincidence may also be explained in terms of the importance that highly educated personnel have for the effectiveness of the 'highly advanced industrialised' economies that make up the Organisation's membership. Since the late eighties, universities have assumed an even greater centrality with the rise of a global economy that is knowledge based and with the new consensus in educational policy which was described in Chapter 2. The OECD has been important in setting out the differing versions of human capital theory which at different times underpinned its approach (Marginson, 1997a, 1997b). As the OECD itself remarked: "In its evolving state, tertiary education is the most dynamic and complex sector of educational policy and practice with ramifications throughout the society and economy" (OECD, 1998c, p. 13). Greater participation by the age cohort in higher education is deemed necessary in view of declining teenage labour markets and to ensure the economic competitiveness of nations. Furthermore, higher education has become a tradeable good within the global economy which may also be a source of benefit to the economies of nations.

The contemporary policy significance of higher education has been evident in the trend towards universal provision and the various policy dilemmas facing nations in attempting to handle this transition. Trow (1974) defined the move from elite to mass higher education as occurring when more than 15% of the age cohort attend, and set universal higher education participation when more than 50% of the relevant age cohort are in attendance. Using this classification, most OECD countries are now moving towards universal participation. Trow unveiled his definition at an OECD conference in 1974; earlier he had produced a document on the 'problems' associated with this transition for the Carnegie Foundation on Higher Education in the USA. The widespread usage of the classification points up the OECD's role as an international mediator of knowledge, classifying and categorising phenomena and providing thereby a language and a terminology for international exchange in the domain of higher education policy.

The OECD sponsored and supported this general move towards universal participation in higher education in a variety of events — conferences, publications, thematic reviews, and through the Programme on Institutional Management in Higher Education (IMHE). During the 1960s the OECD recognised that rapid growth in university participation meant that universities and governments both needed to take more account of "administrative strategies" and "management techniques" inside universities to handle growth and diversity (OECD, 1993c, p. 28). CERI conducted a number of studies between 1969 and 1971 on management of universities which formed the background to a conference in 1971 and this in turn hastened the creation of the Programme on Institutional Management in Higher Education (IMHE) the following year. The IMHE brings together institutional membership — normally individual universities — though the Higher Education Division of Australia's federal Department of Education, Training and Youth Affairs is also a member. University education in Australia, unlike schools, is largely the financial and policy responsibility of the federal government. IMHE represents another network of players beyond the nexus of policy makers, researchers and consultants associated with the Education Committee and CERI. Its changing agendas reflected the multiple pressures upon university management that arose from the move towards universal provision and from reduced public funding. The IMHE Web Page describes its current topics of interest as: "continuing education and lifelong learning, the tertiary education institution as regional actor, diversification of tertiary institutions, management of research, quality in mass higher education, internationalisation of higher education, legal issues and governance" (IMHE, 1999).

Together with almost universal provision, the internationalisation of higher education, often in a marketised policy framework, has served to intensify interest in and preoccupation with management, which in turn justifies the work of IMHE. Marketisation is, moreover, closely related to neo-liberal views on state funding and with pressures on universities to be financially more self-reliant. Simultaneous pressures from the market and new forms of state steering upon the universities, in turn oblige them to reconstitute both their management and internal governance (Schugurensky, 1999). The OECD's report *Financing Higher Education* (OECD, 1990a, p. 10ff) acknowledged the close relationship between funding patterns and the internal organisation, management and governance of universities. Pressures such as these call for university management of a particular kind, which is entrepreneurial and strategic. IMHE has concentrated on such issues, as has the Education Committee, a development noted in the Thematic Review. Indeed, the Review's concluding chapter was very clear on this matter, "The transformations in demand and expectations now occurring and likely to continue call for highly-sensitive and intelligent leadership at all levels" (OECD, 1998c, p. 101).

Conferences have been important vehicles in the OECD's support for debates on various policy options in the field of higher education. OECD Conferences at different times reflect the particular issues raised and which are mirrored in the changes in nomenclature employed. The OECD Background Report (OECD, 1993c) prepared for the conference *Transition from Elite to Mass Higher Education* held at Sydney in June 1993 refers back to two influential earlier conferences. The first, held in 1973, was entitled *Future Structures of Post-secondary Education*. The second, *Policies for Higher Education*, was held in 1981. The former event was imbued with optimism as the sector continued to expand, while the sector itself was somewhat less sanguine about its future condition and public funding when the 1981 conference took place. The 1973

Conference alluded to "post-secondary education", its 1981 successor to "higher education". In 1987, the Report, *Universities under Scrutiny* gave voice to what had become the "beleaguered condition of higher education" (OECD, 1993c, p. 11). Here are revealed the problems of nomenclature — the concurrent usage of 'post-secondary' and 'higher education'. The Background Report for the Sydney Conference in 1993 invoked the concept of 'higher education' as a means to draw attention to definitional and policy issues still unresolved. They included "not only the internal hierarchy and the interrelationships of institutions, programmes and courses beyond the stage of upper secondary schooling, but also ambiguities in the relationships between vocational and adult/recurring education and the higher education sector" (OECD, 1993c, p. 3).

The more recent Thematic Review, *Redefining Tertiary Education* (OECD, 1998c), brought in yet another category, that of 'tertiary education'. Tertiary education, the Thematic Review stated "refers to a stage or level, beyond secondary and including both university and non-university types of institutions and programmes" (OECD, 1998c, p. 9). Taken together — universities and vocational education and training — may accordingly be described as 'post-secondary', or as in the terms set down in the Thematic Review, as 'tertiary' education. The Review justifies this change in terminology on the grounds that with the transition to universal participation in higher education, tertiary education to all intents and purposes becomes taken-for-granted after primary and secondary schooling. This chapter focusses upon universities alone — hence its title 'Redefining University Education'. Chapter 6 dealt with the other component of post-secondary or tertiary education, namely vocational education and training, which when set in relation to the transition to work, was the subject of the other Thematic Review the OECD conducted.

From the outset, the OECD's involvement in higher education reflected the issues of policy associated with the move from elite to mass and on towards universal higher education. While OECD's initial interest in management of universities followed the expansion of the 1960s and 1970s, the transition during 1990s to universal provision of university education amongst its members has deepened and widened such concerns within the global context and neo-liberal policy framework. If the chronology briefly alluded to earlier is brought into play the sustained nature of these issues is fully apparent. Certainly, a line may be drawn between concern with the transition from elite to mass tertiary education in 1974 and the recent Thematic Review. However, contemporary anxiety over the effects of greater participation is seated in a very different political and policy context. The 1974 Conference *Problems in the Transition from Elite to Mass Higher Education* outlined the rapid change in participation rates against a background of confidence in the role of universities. In stark contrast to this the Report, *Policies for Higher Education in the 1980s* (OECD, 1983b), which emerged from an intergovernmental conference in October, 1981, dwelt long on the *lack* of public confidence in universities. The 1987 Report, *Universities under Scrutiny* (OECD, 1987d), written by the Vice-Chancellor of the University of Hull (England), Professor William Taylor, with a Secretariat reply appended, raised issues of purpose, of links with economy and society, funding approaches, and pressure for greater efficiencies. *Financing Higher Education: Current Patterns* (OECD, 1990a) specifically addressed funding options for universities given the restraint on government expenditure and given too the growing place of the market. These three reports examined university policy set against the looming globalisation of the economy, the rise of neo-liberal policies and the accompanying (government) enthusiasm for parsimony in the funding of higher education. *Redefining Tertiary Education* evaluated the policies in place from that time and seems to be more assertive in proposing a way through the resultant dilemmas. Furthermore, the report set the construction of its context more firmly in terms of globalisation. At one level, it attempted to deal with the 'incipient dangers' that faced all university systems in OECD countries as outlined in *Universities under Scrutiny*, amongst which were the role of universities set against greater secondary school completion; the purposes and design of undergraduate curricula; the place of research in expanded university systems; and questions related to funding and management.

The OECD from time to time conducted reviews of the educational systems of its member nations. The exception to this was the 1987 review of the Californian higher education system. However, country reviews have sometimes focussed on universities; for example, a 1971 exercise examined the role of the US federal government in research and development, while the 1971 Japan country review was largely taken up with higher education. Country reviews were undertaken at the request of the member nation. Recently, however, members appear to have lost their enthusiasm for such events. With the Cold War ending, the OECD has reviewed more aspects of the educational systems of 'economies in transition' than reviews of member countries; reviews have been made of Hungary, Russia, and the Czech and Slovak Republics and link to the OECD's outreach, extending its sphere of (policy) influence.

In this setting, the *Redefining Tertiary Education* Thematic Review is important. It is the first cross-national review of its kind to be undertaken by the OECD and, not least, because it may herald a new *modus operandi* for the Organisation. Certainly, the Review points to the emergence of a field of policy dealing with tertiary education in a global dimension. Universities, in their cosmopolitan as opposed to local guises, have always been global in their knowledge, their research approaches and in the circulation of staff and ideas. However, globalisation entails a qualitative shift in these global flows — advanced technologies annihilate the restrictions of time and space — which, in a phenomenological sense, constructs the world as one place. Purposes and functions, operation and management of universities, together with relationships to government and markets, all have been profoundly affected. And as we have consistently argued, the OECD has played a significant role in articulating and spreading abroad responses pervasive in their neo-liberalism and new managerialism which meet these emergent conditions. The Thematic Review analyses both changes and the policy directions tertiary education may take in the future, emphasising the first years and meeting the needs of students. Yet, it does not mention explicitly the overall significance of the OECD as promoting both the ideologies of neo-liberalism and of new managerialism.

The Thematic Review inaugurated in 1995, used the format well tried by National Reports — preparation of a country report, reviewers' report, the examination meeting, and publication of final report — with some modifications incorporated. While National Reports were implicitly comparative, the Thematic Review from the outset was explicitly comparative. The ten participating countries were Australia, Belgium (Flemmish community), Denmark, Germany, Japan, New Zealand, Norway, Sweden, United Kingdom, and the United States (Commonwealth of Virginia). They replied to a common set of questions and issues and were visited by a review team. All had recently experienced comprehensive reform in tertiary education or were contemplating it. The Secretariat, after wide consultation, drew up an overview of the Review Team's findings. A wide range of data was drawn from official statistics, from research and policy literature. The Education Committee and an advisory group participated throughout the exercises. The review teams visited the ten countries from eight to ten days from June 1995 to October 1996, each team consisting of two members of OECD's Secretariat and two independent experts.

In the first chapter, the Thematic Review outlined its goal: to determine how far the first years of tertiary teaching were adequate to the needs of students, of the economy and of society (OECD, 1998c, p. 17). Additionally, it sought to "analyse how future policies might best promote needed change" (p. 17). The review defined its task as: to document significant trends in growth in participation in tertiary education amongst member countries; "to develop new perspectives and concepts to inform and strengthen policy analyses" (p. 9); to ascertain key aspects of change "in which a co-operative international review could facilitate policy development"; and "to explore new methods of undertaking OECD education policy reviews by defining a common theme of concern" (p. 9). Thus it gave voice to goals and purposes as they applied to teaching in the first years of universal tertiary education in ten member countries and as they related to its own *modus operandi* in a new geo-political context. It is to a closer analysis of the Thematic Review that the chapter now turns.

The Policy Themes of Redefining Tertiary Education

In this section, the major themes of the Review will be briefly sketched and the tentative policy put forward as a solution to a number of dilemmas will also be considered. While the report is client focussed, that is, it focusses on students in initial qualification courses, a large number of broader issues concerned with tertiary education are taken up — amongst them student demand, contending rationales beneath the policy of tertiary education, curriculum, teaching and learning, quality issues, leadership, management, costs and financing.

Growth in demand for tertiary education has been substantial across the countries examined. Though some minor fluctuations are visible, the prediction is that demand will continue. Universities have to change to meet the needs of this new student body, the Report argues. If this situation may be "disturbing", it provides "new opportunities" (p. 22). The capability or preparedness of entering students stand foremost, as does the role of post schooling vocational education which, the Review asserts, has become as much a remedial bridging preparation for tertiary participation as a goal in its own right. The issue of success as opposed to access is raised. The place of research in mass tertiary institutions becomes an issue, as does the question of quality. Finland, the report observes, seeks to have sixty to sixty-five percent of the age cohort in tertiary education by the early 2000s without a diminution of quality.

Enhanced participation in tertiary education stands as a bulwark against high levels of youth unemployment which in turn is seen "to impede growth" and create "social risks" (p. 38). Certainly, employment for graduates has changed over time, from the professions to public sector employment and more recently to private sector work in commerce and financial services. There has been something of a symbiotic relationship between the decline of the public sector and of public sector employment for graduates under neo-liberal policy frameworks and the rise of the more market driven universities producing entrepreneurial individuals for the private sector (Neave, 1997). Institutions need to take account of these changes and, the Review suggests, they should work to meet the needs of that other client, the employers of graduates. It is suggested that the generic skills necessary to meet these new employment demands and to promote the capacity to change careers over a life time are "adaptability, creativity, initiative and team work" (p. 39).

Diversity in institutional provision is required to meet the needs of the more heterogeneous student body associated with universal participation. However, some factors inhibit diversity of course and institutional provision, so varying types are to be found in both binary (e.g. Germany) and unified systems (e.g. Australia and the UK). The Review notes:

> "There are also limits to diversity: government policies, for example; the internationalisation of science and culture; concepts such as generic skills and consistent standards; the benchmarks of a liberal education; and not least, the relative homogeneity of background of many tertiary teachers."

> (OECD, 1998c, p. 41).

The Thematic Review sketches out in some detail the competing rationales for universal tertiary education which split between the ethic of "utility in the market place" as opposed to "the advancement of knowledge". This dichotomy touches upon other aspects of universities, their margin of autonomy from society and the economy in setting research agendas and course content. It also bears upon how universities are run, either in a collegial or in a managerial fashion. Whilst endorsing a convergence between the utilitarian and critical intellectual traditions, the Review recognises the broad social consequences of universal provision, including enhanced cultural tolerance, greater social justice, better quality of life, over and above the economic benefits that flow from a "self-reliant, innovative and competent workforce" (p. 44). The Review proposes reconciling these competing traditions by focussing on the needs of students in the first years, by on-going analyses of social and economic changes, and by taking account of the needs of employers and professions.

Disciplinary boundaries are changing and new knowledge domains emerging. The question is asked: "Need the functions of knowledge advancement, practical utility and sale of services be seen as contradictory?" (p. 46), with an implied denial. The report backs both collegial and managerial approaches. It supports greater representation of industry and the professions on university boards and governing councils. It calls for more partnerships, with industry, and with regional and local communities. There is also a plea for much better coordination of courses across institutional types and across courses. If decisions are to be devolved to individual universities, greater institutional cooperation is demanded to improve the liaison, credit transfer and recognition of prior learning. Such cooperation is blunted by a competitive market approach to tertiary educational policy.

The Review considers curriculum design and teaching in the universities (Chapter 5). The question of control of curriculum is canvassed, with some recognition that supranational bodies (for example the European Union) might have some impact. It also suggests that in a more globalised future, the World Trade Organisation could become involved in the cross-national recognition of professional qualifications (p. 56). It is also recognised that the professions, industry and students as consumers are having a bigger impact upon curricula than heretofore. This situation re-engages the conflict of the university acting in the traditional pursuit of knowledge as opposed to a mercantile interpretation of its purpose — a dichotomy the Review suggests ought to be eschewed. Though the Review agrees that status and promotion in universities are commonly associated with research and publication, it also encourages universities to be more interventionist in respect of teaching, through both quality and accountability mechanisms and career reward structures.

Quality has long fascinated the OECD, which focussed on quality in schooling during the early eighties and moved later to tertiary education. IMHE confessed to an ongoing interest in quality issues in tertiary education as well. The Review sets the issue of the quality against the enhanced entrepreneurial activities of universities and against the drive towards increased participation. Quality in education is a controversial concept with multiple meanings and applications. The Review stresses three elements, attrition rates amongst students (it called for more and better data on this phenomenon), systemic evaluation of quality, and leadership and quality. The issue of standards is also evoked though universal participation renders it unrealistic to accept a definition akin to earlier more elite provision. Introducing teaching evaluations is applauded, so is the inclusion of teaching in determining tenure and promotion. The culture of quality audits, it notes, while sitting uneasily with some in academe, has been accepted quite rapidly into the operations of universities and university systems. Recognition is also bestowed upon the heavy bureaucratic demands national quality audits make upon individual institutions, quite apart from the usefulness of much of the data collected. Here the culture of performativity considered earlier (see Chapter 5) raises its head. Tensions emerge from steering through state policies and at the same time through the market. Quality procedures, when they engender league tables also play into the global marketisation of universities, an issue we will return to later. League tables constrain diversity of provision through their implicit subscribing to the assumption that there is one model of the 'ideal' university.

The Review documents the gathering momentum in the higher education systems of its member states towards incorporating the "principles of private sector business management with heightened executive responsibility and authority and sharper forms of accountability than hitherto" (p. 77). Such progress has set the stamp on a cultural and occupational divide between executive positions and the "main body of teachers and researchers" (p. 77), eroding collegiality. The Review admits that tension exists between efficiency and equity concerns when dealing with student selection and attrition rates. As it does with the dichotomy of market utility versus the production of knowledge, here too the Review seeks to hew a delicate path between a collegial or managerial mode of leadership within institutions. Different relationships between system governance and institutional governance are also touched upon in an attempt to reconcile both system accountability and responsiveness at the local site. Leadership, it is argued, should take account of academic cultures and histories. The stance finally hit upon believes that different approaches are best applied to different types of decision-making, for example, collegial for academic decisions and managerial for strategic, institutional planning decisions. Through appropriate leadership, the Review anticipates that:

> " ... the difficulties can be overcome of reconciling 'collegiality' and 'managerial' styles, of reconciling highly local, regional, national and international perspectives, of distributing decision-making and enhancing the sense of responsibility of all the stakeholders towards the clients."

> (OECD, 1998c, p. 83)

The costs of financing universal participation in tertiary education are considered in some detail. The possible negative effects of a fall in per-student expenditure at the same time as increases in student numbers are also contemplated. An investment perspective (human capital theory) on university expenditure is most often accepted by both governments and individuals. (However, the rise in student contributions suggests that governments see university expenditure as a cost.) Both economic and individual benefits accrue from participation in tertiary education. How different systems have tried to fund increased participation in tertiary education are explored. The report believes that greater efficiency is necessary in the usage of current funds, but alternative sources of funds are also required. Certainly, institutions have been urged to seek funds from sources other than government. Equity considerations are also present here. There does not appear to be a direct correlation between fee structures, levels of participation and equity across the various systems that the Review investigated. Agreed, change is easier when well lubricated by extra funds (p. 93). And indeed "scarcity creates its own creativity" (p. 93). The Review's stance is that the market is not a panacea to the problems of funding universal tertiary education or of effectively developing policy.

Analysis of Redefining Tertiary Education as a Policy Text

The Thematic Review is an excellent example of the way policy documents create their own contexts as well as responding to them (Seddon, 1994). The Foreword sets the context for the changes to higher education on which the Review focuses: "The changes are set in a new context, one in which much greater value is being placed on the skills and flexibility of individuals as a key to reducing unemployment" (OECD, 1998c, p. 3). This observation gives rise to the interpretation that lack of skills and flexibility are the cause of structural unemployment. Thus, the document creates its problem context and throughout the Review proffers a solution, namely, universal tertiary education of a particular kind. The Foreword then notes that these very skills and flexibility are also the key to "improving economic performance" (OECD, 1998c, p. 3). Thus the enunciation of a problem — poor economic performance — anticipates throughout the Review, the implied solution — universal participation in tertiary education. The Foreword ties this problem setting and solution to a current major theme of the Organisation's work, "a lifelong approach to learning" (p. 3). Not all stakeholders in higher education agree with the Review's conceptualisation, yet, the Foreword claims that a "sweeping shift in orientation" is occurring:

> "While all might not agree with the details of the vision, the forces at play suggest a sweeping shift in orientation toward even higher levels of participation at the tertiary level, driven strongly by demands reflecting the diverse interest of clients rather than supply-led, institution-directed expansion witnessed previously."
>
> (OECD, 1998c, p. 3)

Out of the process of problem construction and problem framing, the Foreword sets up a number of challenges which universities and university systems must face. Firstly, how tertiary education can better respond to client interest. Secondly, how may the needs and interests of those in the first years of tertiary education be met better. The needs of first-year students become crucial given the greater diversity in the student body and given the participation of hitherto previously under-represented groups. The language of inclusion serves to address this question, dovetailing with the emergent language of policy in the European

Union and many European states. Funding tertiary education then becomes paramount especially when neo-liberal policy is committed to paying back state funding. The final topic developed in the Foreword of the Review deals with the type of approaches to the steering of government policy that may apply to a differentiated system and demands for institutional autonomy. The Foreword puts forward the idea that the Thematic Review may even mark "a turning point" in higher education policy.

> "It will serve as well as a basis for debate, reflection and exchange as all parties seek to maintain and extend the contribution of tertiary education to the improvement of the economic and social well-being."
>
> (OECD, 1998c, p. 4)

Policy always involves a struggle over meaning or a "politics of discourse" (Yeatman, 1990). The Review focusses specifically upon students — deemed 'clients' — a term in itself indicative of the discursive reconstruction of the university in an era of neo-liberal policies and marketisation. If the Review accepts that the "client" is the "dominant motif of this inquiry" (OECD, 1998c, p. 15), it equally admits, that in some quarters of the university policy field, this language appears as "a perverse concession to the consumer society and a failure to acknowledge the independence of knowledge and critical inquiry" (p. 15). Having made this passing concession, however, the Review then reverts to its basic thesis:

> "But the client perspective, in challenging many of the orthodoxies and pre-suppositions of institutional life is on the one hand consistent with demand-driven policies and structures in our societies and, on the other, emphatic that there are new needs to meet as we move towards universal tertiary education."
>
> (OECD, 1998c, pp. 15–16)

Policies as texts are then framed and constrained by overarching discourses (Ball, 1994). In this instance the OECD indicated its acceptance of the new imperatives of demand side tertiary education and the need for change. The implicit or not so implicit normative stance is revealed — an admixture of the apparently academic, distanced, and disinterested with the normative — an approach redolent of the genre of OECD reports. The Thematic Review even admits this when it presents itself as "descriptive, analytic and normative" in orientation (OECD, 1998c, p. 18). It explains that the Review has been undertaken "to allow members to take the next steps in policy formation" (p. 17), implicitly suggesting an appropriate direction for such steps. Likewise, the argument is made that "the success of the review will be judged by the clarity it brings to the issues, and the insights and stimulus it provides to future policy and action" (p. 17).

Throughout the Review the need to develop what might be appropriately called hybrid universities, everywhere in evidence. Hybrid universities reflect the new demands of clients and other stakeholders in a globalised world while retaining older conceptions of the university. Such hybrid institutions would have to amalgamate these new demands with older scholarly concerns of curiosity-driven research, the dispassionate search for knowledge, institutional autonomy, academic freedom and social critique. The new hybrids would support the production of both social and human capital. Here, the normative aspect creates a conceptual consensus beyond an 'either/or' model of the old and new university. Critics of current developments in universities often attack these demand and market side interpretations and hark back to earlier approaches, to the Humboldtian and Newman ideals, to those of the sixties, where in hindsight Clark Kerr's 'multiversity' begins to look even attractive (Miyoshi, 1998; Schugurensky, 1999). *Redefining Tertiary Education* seeks to cut through this debate by embracing an 'and/also', rather than 'either/or' approach to the conception of a university. In so doing, it re-states the concept of a multiversity set within a global framework of new governance, market economy, and parsimonious public funding, along with universal participation.

Redefining Tertiary Education is a good example of both the policy genre and the OECD report genre. It is a 'greedy' report, not wanting to eschew any of the traditional stances of university education, management and funding from the era of elite provision. At the same time, it supports the new entrepreneurial, marketised university and universal provision. It also circumvents an either/or stance with respect to the rationales of market utility/advancement of knowledge to justify university education and managerial/collegial approaches in running them. The Review thus seeks to pull in all sides of the debate, whilst denying many of the realities that result from the new framings of policy in universities. This purblindness reflects the ideological framework in which the Organisation is situated, the tensions in which were explored in detail in Chapter 4. Some of the difficulties with the Review's and/also stance are considered in the following section, which also suggests that the normative stance anticipates an emergent role for the OECD as policy actor.

Redefining Tertiary Education: The OECD as Policy Actor?

Redefining Tertiary Education provides a typical example of the OECD report genre, seemingly distanced and objective while being implicitly normative. In this instance, the implicitly normative accepts neo-liberalism and a micro-economically focussed theory of human capital, while remaining silent about the OECD's role as advocate of these two approaches. The OECD thus becomes a policy actor setting out a way forward for tertiary education which flows back into national and institutional dynamics of policy production and reinforces developments there. The Review recognises (and anticipates) convergence in cross-national educational policy, but as is its wont concedes that different countries will pursue these agendas in varying ways (OECD, 1998c, p. 107).

As Appadurai (1996) noted, the global, national and local play off each other in creative and interesting ways, resulting in what Stephen Ball (1998) has so felicitously characterised as "big policies in a small world". An interplay exists between "context-productive" (global, supranational, national, top-down) and the "context-generative" (localised, vernacular, bottom-up) policies and practices (Appadurai, 1996). This gives rise to specific local outcomes, in national higher education systems and in particular universities that are still, nonetheless, effects of globalisation and supranational policy discourses such as those espoused by the OECD (Lingard, 1999). Commenting about the globalisation of the economy, Simon Marginson (1997a) has noted how global and national systems of policy development reinforce each other.

> "Globalisation weakened the capacity of national governments to determine the content of national policy agendas, now often steered from a global distance. But globalisation also empowered national authorities, who became the interpreters of global agencies and market requirements, and the arbiters of local reform. The two systems of control, national and global, reinforced each other."
>
> (Marginson, 1997a, p. 84)

Marginson's observation about economic policy seems to apply quite aptly to the situation in university education.

The critics of contemporary university policies, for example Schugurensky (1999), see very real tensions and contradictions between the steering dynamics of the commercial model and state interventionism, as well as challenges to university autonomy that flow from these two competing dynamics. Drawing on Weber, Schugurensky (1999, p. 297) argues that university autonomy has been replaced by a heteronomous approach. Under the emergent heteronomous model, missions, agendas and outcomes are designed and framed by both market dynamics and state pressures. Dealing with this context, the Review seeks to synthesise the notion of education as a good for private consumption with the tradition of the disinterested pursuit of knowledge in autonomous institutions. In contrast, Schugurensky (1999) argues that the former tends to crowd out the latter so that heteronomy replaces autonomy:

> "Education is considered more as a private consumption or investment than an inalienable right or search for disinterested knowledge. As a language that centres on user fees, rational choices, job prospects, and private rates of return becomes increasingly hegemonic and the commercial model becomes paradigmatic, attacks on competing models escalate. The 'academic haven' model (scholars seeking truth in an uncontaminated environment) is perceived as an irrelevant ivory tower."
>
> (Schugurensky, 1999, p. 298)

Equity is rightly a concern of the Review, but its approach appears to be a contingent one — greater participation will mean more equity. The evidence on the ground would seem to suggest a new hierarchy of institutions is emerging. Those at the top educate the new mobile elite who participate in global labour markets, with lower status institutions producing others for less prestigious careers grounded in the local economy. Some of the latter are involved with 'employability' and keeping people out of the labour market for longer periods (Young, 1998). Finally, there is the issue of those excluded from tertiary education altogether, who end up as the losers of globalisation. Universal tertiary education with its hierarchy of inclusions and exclusions contributes to occupational and territorial mobility/immobility, which, Bauman (1998) argued, represents a new dimension of inequality in globalised labour markets.

Tertiary education, the Review claims, is as much about producing social capital (without using the term) as it is about producing human capital. Schugurensky (1999, p. 299) argues that the heteronomous model with both state and market steering may weaken the social mission of the university. Furthermore, social critique can be marginalised when institutions depend on the corporate dollar (Miyoshi, 1998). According to Miyoshi (1998, p. 263): "Global corporate operations now subordinate state functions, and in the name of competition, productivity, and freedom, public space is being remarkably reduced". In such a setting, the universities may be less inclined to offer critiques of corporate and state policies and less inclined to work in the interests of the common good (Miyoshi, 1998). *Redefining Tertiary Education* slides over some of these difficulties by calling for greater resource efficiency in universities and the mobilisation of resources from many sources, including the state and the private sector to fund universal tertiary education.

In its conclusion, *Redefining Tertiary Education* develops the idea that rapid changes in knowledge production and in labour markets and career structures not only require more university graduates, but more graduates of a particular kind. The Conclusion thus states: "tertiary education should be developing wider capabilities, more entrepreneurship and a more flexible attitude towards working life" (OECD, 1998c, p. 105). It also calls for more attention to "cross-curricular and generic competences" and greater provision of "more experiential and problem-based learning" (p. 105). The use of entrepreneurship is of more than passing interest. The Review presents the dominance of neo-liberal policies as "the rebirth of enterprise culture", but also admits that employment insecurity follows in its wake (p. 38). These usages appear to link up the way in which the global pervasiveness of neo-liberal economic theory has ground down the distinction between the economic and the social dimensions with individuals now defined in terms of their ties with the economic (Rose, 1996). Often, the end of educational policy is held to be employability pursued through lifelong learning, with the onus on the individual rather than on the government to create jobs. In this context, educational systems increasingly are taken up with producing a particular entrepreneurial self: a self responsible for the creation of a 'portfolio career' and for its own well-being. Here is the new form of governance in a global world. The Conclusion to *Redefining Tertiary Education* begins to head in this direction, despite all of the rhetoric lavished on the goals of social, democratic and inclusion injected into this level of education. Neo-liberal policies "re-figure" the territory of government both in their individual and economic fetishised foci and their elimination of the social dimension (Rose, 1996).

Schugurensky (1999) analysing "Higher Education Restructuring in the Era of Globalization" deals with the same set of issues considered by *Redefining Tertiary Education*. Indeed, the dilemmas facing universities which he raised in his concluding comments echo quite closely those addressed by the Review:

> " ... the main challenges for universities are how to contribute to economic development while preserving integrity, autonomy, and community interests, how to balance an efficient management with democratic governance, how to expand while protecting quality, and how to engage in scientific and technological ventures that are guided by human and ethical values."

> (Schugurensky, 1999, p. 301)

As the OECD has also recognised (1993c, p. 66), universal participation in tertiary education brings many pressures to bear upon the traditional purposes of universities, and challenges the Humboldtian and Newman traditions of an intimate and necessary link between research and teaching within the functioning of universities (OECD, 1998c, p. 57). Universal participation has also precipitated a more utilitarian challenge to their liberal and humanist educational goals. It is a change which raised questions to do with university teaching to a student body that comes from a more diverse range of social and educational backgrounds than during the era of elite provision. Questions of university teaching also link to considerations of curricula as the student body and institutions also become more differentiated. The Thematic Review deals well with questions of university teaching in the era of mass provision. But it glosses over questions of the relative place of teaching and research in new universities dependent upon entrepreneurial consultancies and marked by differentiated statuses and labour markets.

The Review regards greater diversity of provision as essential to meet the needs of a larger and more diverse student cohort. It also recognises limits to such diversity. That the university of the future might well be both comprehensive and differentiated is another possibility (p. 41). A recent study of diversity in the unified national system of universities in Australia analysed the degree of diversity, both systemic and programmatic, within that system, and found it wanting in both aspects (Kemmis et al., 1999). It argues, as does the Thematic Review, that more diversity is necessary not simply because of greater diversity in the student body and for reasons of equity, but also because of changing social relations and the new global knowledge economy. In terms of equity it supports an interpretation of 'justice-as-difference' and rejects a straightforward distributional definition. Kemmis et al. (1999) show how the increasingly quasi-market approach to policy at system level in Australia curtailed diversity and instead encouraged isomorphism and mimicry. Even in a more marketised setting, the character of education as a positional good poses little challenge to elite institutions, still less to the dominant model of university education they are held to provide.

Kemmis et al. suggest that diversity should be conceived in terms of the core business of universities, on the lines of the thesis by Boyer (1990) of the four dimensions of scholarship, that is, teaching, application, integration (synthesis) and discovery. Each connects with different fields and clients. Through their practices of teaching, universities connect with students and graduates; through practices of integration, they connect to the community and public sphere; through practices of application, they relate to the professions, business, industry and government; and finally, through discovery, they connect to communities of scholars and to different intellectual fields (Kemmis et al., 1999, p. 27). These four practices of scholarship provide a useful way forward in conceptualising diversity in tertiary education. Equally relevant, Kemmis et al. deal with the issues broached in the Thematic Review, which seems to hint at some universe of discourse about contemporary tertiary education policy. However, their view would seem to suggest that the stance taken by the Thematic Review vis à vis diversity in a neo-liberal policy framework is somewhat bowdlerised. In fairness, it has to be said that the OECD Review concedes that an emphasis upon research, the effects of technology and the culture of internationalisation all serve to inhibit the diversity required (p. 42). Even so, it fails to recognise the tensions between market and diversity, between old elite institutions and newer ones. The Thematic Review's invocation of diversity in some ways glides over these difficulties, many of which are tackled by Kemmis et al. Thus the same topic and similar themes produce very differing degrees of recognition and confusion over the possible inhibitors to diversity within present-day university systems.

Worthwhile and intelligent though the analysis provided by the Review is, it glosses over the tensions and the dilemmas which universities face in an era of universal provision and globalisation. As a policy solution it espouses an 'and/also' approach in encouraging new hybrid tertiary institutions. Furthermore, it either fails to recognise, or misidentifies, the depredations of globalisation read as neo-liberal economic reform. It could hardly do otherwise, given the important role the OECD has played in pursuing neo-liberal market reform, and its organisational fellow, corporate managerialism. In stark contrast, Schugurensky (1999) locates the massive changes affecting universities within this new global world and its related "economic, ideological, and political forces". Amongst these he includes: "the globalisation of the economy, the implementation of neo-conservative and neo-liberal policies, the consolidation of international corporate powers, and a redefinition of the role of the state" (p. 299).

The OECD is far more sanguine about these forces than either Schugurensky or for that matter, ourselves. Given the ideological tensions within the OECD, explored earlier and at length (see Chapter 4), this optimism is probably unavoidable. As it continues its role as policy forum, the OECD has taken on the (surreptitious) role of policy actor through its discursive interventions as an enthusiastic supporter of globalisation read as neo-liberalism, of new managerialism, and of anticipated policy convergence in globalised and internationalised universities and university systems. Convergence and standardisation of this nature are effects of such global institutions as the OECD, as John Meyer and his colleagues have argued (Thomas et al., 1987; Meyer et al., 1992a, 1992b, 1997). It is to a consideration of the OECD and the internationalisation of university education that this chapter now turns.

The Internationalisation of Tertiary Education

The preceding analysis of *Redefining Tertiary Education* illustrates the ways in which the OECD is able to occupy an ambiguous policy zone, both as a forum, designed to facilitate intergovernmental dialogue and as an actor, an active prosecutor of particular policy preferences. Able to appear impartial and objective, it is at the same time working to bond together a policy settlement in keeping with its overriding support for neo-liberal market reform and corporate managerialism. There are few better examples of the politics of ambiguity than the OECD's promoting of the idea of the internationalisation of university education. In its mature form, internationalisation may be viewed as a concept that embodies the Organisation's ideological preferences. It is a major plank in its strategy for bringing about structural change in university education. By means of internationalisation, universities are expected to become more responsive to the OECD's interpretation of economic globalisation.

In the development and promoting of its normative agenda, the OECD has held a number of important conferences devoted to the subject of internation-alisation of university education. It also published key reports, designed to encourage universities to develop a global perspective in their work as a neces-sary response to the emerging global environment. They ensure member country universities and their graduates to remain competitive in the international market. The OECD has produced guidelines for approaches to the internationalisation of university education. It has set down myriad examples of 'best practice' that pervade the discourses of international education across much of the world. A chronological analysis of these guidelines reveals the OECD as managing a process of policy production, rather than merely providing a forum where ideas are explained. Its discursive interventions are designed to paper over policy differences, to forge a synthesis out of competing positions that nonetheless conform to its dominant educational ideologies. Nowhere is the evidence of this synthesis better found than in the OECD's attempts to develop a range of indicators to measure the degree to which universities have internationalised.

As Haarlov (1997), an influential Danish scholar closely linked to the OECD, argued:

" ... internationalisation has always been part of the higher education institutions's activities, both within research and increasingly within education. There is a growing incentive to have an international dimension included in higher education programs, partly because of labour market stipulations to this effect and partly because social developments in general are heading towards a multicultural and more globally minded society."

(1997, p. 26)

Haarlov adds, however, that "the internationalisation of education has taken on new dimensions in the past ten years, both in the form of increased student and teacher mobility, and in the form of a more international approach to the specialised content of many programs".

Haarlov's views represent a distinctively European perspective on the internationalisation of university education. Its rationale is broadly in line with the European Union's Erasmus Program that highlights student mobility. This perspective stands apart from recent formulations of internationalisation in Australia, Canada and the UK, where commercial concerns have been to the fore. In the early 1990s, at least three perspectives on internationalisation could be found within the OECD. They reflected the distinctive manifestations of internationalisation in the regions of North America, Europe, and Asia. Where the North American view emphasised commercial aspects of internationalisation and the need to establish international contact across geographic borders, the European notion of international curriculum reflected Eurocentrism in response to European Union preoccupations at the time (e.g. the eventual integration of Eastern Europe into the European Union). Asia, on the other hand, was undergoing the increasing movement of foreign students into the region, together with continuing flows of local students to international destinations. From these various conceptions, the OECD has developed its own position. Its position suggests that internationalisation is concerned not only with the international exchange of students (which constituted the main thrust of internationalisation at the time). It ought also to focus on changing the substance and delivery of curricula for domestic students (OECD/CERI, 1996). Both market and cultural views of internationalisation have been accommodated side by side in some degree of tension each with the other.

In educational terms, the OECD's perspective on internationalisation is a hybrid combination of reforms. These may variously involve academic and extra-curricula activities, the development of new skills, attitudes and knowledge in students, faculty and staff, the development of an ethos, which values intercultural and international perspectives, and/or the injection of an international dimension or perspective into the major functions of the learning institution (Knight and de Wit, 1995). Jane Knight and Hans de Wit grouped the overlapping rationales and incentives for internationalising higher education into two broad areas of justification:

(1) *Economic and political rationales*, such as economic growth and investment in the economy of the future; to be competitive with the international labour market; to foster diplomacy through educational cooperation; financial incentives (e.g. contract education, recruitment of foreign students and international education advisory services to generate income); the national demand for higher education is so great that nations encourage study abroad.

(2) *Cultural and educational rationales*, either to export national, cultural and moral values, or to increase intercultural knowledge, skills and research; to expand social learning and development of the individual; to provide

an international dimension to research and teaching; to strengthen the core structures and activities of higher learning institutions through international cooperation; to improve the quality of education and research (Knight and de Wit, 1995).

Knight and de Wit are consultants to the OECD, and may be regarded as the Organisation's main theorists on the internationalisation of higher education. Through a number of reports produced by the OECD, they point out that one of the reasons why there is currently a deficit of coherent strategies for internationalisation of the curriculum in higher education is that no single definition of internationalisation exists (Knight and de Wit, 1995). They insist that a comprehensive understanding of internationalisation must involve a commitment to the development of new skills, attitudes and knowledge in students, faculty and staff. Accordingly, they assert: "Internationalisation of higher education is the process of integrating an international/intercultural dimension into the teaching, research and service of the institution" (Knight and de Wit, 1995).

At the end of the 20th century, policy makers at national levels took up this normative view of internationalisation throughout OECD member countries. The Australian Vice-Chancellors' Committee (AVCC) for example defines

" . . . internationalisation of higher education as 'the complex of processes that gives universities an international dimension'. We thus see internationalisation as relevant to all facets of university life, including scholarship, teaching, research and institutional management. It affects students, staff, and curriculum development."

(Apec, 1997)

For Haarlov (1997), the internationalisation of higher education entails "a complex interplay between history, politics, educational policy, knowledge development and use, and teaching and learning". With improved communication systems, we are all now part of the global village and must both influence and be influenced by it, he maintained. Sharing knowledge internationally is no longer an option, but a necessity. He develops this line of argument further:

"Internationalisation can be characterised as a process of transformation in which areas of activity are increasingly geared to operating in international surroundings, under international market conditions and with an international professional orientation."

(Haarlov, 1997, p. 23)

This thesis has elicited a great deal of work on the ways in which higher education curricula might be internationalised. Perhaps the most influential construct for the internationalisation of curricula was developed during a groundbreaking seminar organised by the OECD's Centre for Educational Research and Innovation (CERI) in 1995. The seminar sought to draw up a comprehensive view of internationalisation by synthesising approaches taken by member countries. Pooled together was a range of ideas, and as is common with such syntheses, the resultant definition turned out to be as general as it was uncontroversial. Prevailing political understandings and ideological leanings were upheld. International curricula were defined as:

> "Curricula with an international orientation in content, aimed at preparing students for performing (professionally/socially) in an international and multicultural context, and designed for domestic students as well as foreign students."
>
> (OECD/CERI, 1994, p. 9)

As this broad definition suggests, curricula may be internationalised in a number of ways. Consequently OECD/CERI developed a typology of characteristics of international curricula, which included:

(1) curricula with international content (e.g. international relations);
(2) curricula that add a comparative dimension to traditional content;
(3) career-oriented curricula;
(4) curricula that address cross-cultural skills;
(5) interdisciplinary programs such as region and area studies covering more than one country (e.g. Asian Studies);
(6) curricula leading to internationally recognised professions (e.g. international business management);
(7) curricula leading to joint or double degrees;
(8) curricula whose parts are offered at off-shore institutions by local faculty; and
(9) curricula designed exclusively for foreign students. (OECD/CERI, 1996)

In their elaboration of the framework provided by CERI, Bremer and van der Wende (1995) provide a more comprehensive analysis of what constitutes "internationalised curricula". In their critical review of *Internationalising the Curriculum in Higher Education: Experiences in the Netherlands* (1995), these two European scholars drew upon the typology of internationalised curricula to illustrate the variety of programs that may be considered in planning for an internationalised curriculum.

The dimensions of internationalisation listed in the OECD/CERI typology often overlap, so a given course of study will often fit into more than one category. For this reason, the typology has been criticised for classifying international curricula into categories "that are not discrete enough to be meaningful". Mestenhauser (1997, p. 7) cited as an example "the category of curriculum that used comparative method transcends all other categories, because everything in international education is by definition 'comparative'; we always study things in relationship to other countries even when we do not say so specifically". His criticism of this 'blurry' dimension is not entirely justified. His suggestion that a comparative orientation is intrinsic to international education does not account for those instances when an international curriculum is adopted in unplanned and/or uncritical ways. Much of the internationalisation taking place in nations of the Global South, Indonesia being one, is unsystematic and has been shaped by the cultural imperialism of education providers from the North (Cannon,

1997). Instances such as this show the desirability of locating and understanding the *rationales* for internationalisation of the curriculum.

One final point about international curriculum deserves further attention: the distinction between curriculum content and form. International content is defined as incorporating an international comparative element or module in the study program (e.g. international politics). International curriculum development may vary in style, and may take the form of single courses or modules, Master's or Bachelor's programs developed within individual institutions, joint curricula or double degrees developed in professional cooperation with institutions in other countries. The Socrates program, for example, encouraged internationalisation of curricula in European Higher Education through international cooperation to develop joint curricula.

In Australia, the content of internationalising curriculum is associated with a combination of local and international values — openness; tolerance; cosmopolitan experiences; and appreciation of global cultural diversity (Cope and Kalantzis, 1998). Beazley (1992) suggested that internationalisation "involves making courses and teaching methods more internationally competitive through links with business and through agreements with foreign governments and educational institutions". Agreements such as these attempt to consolidate reciprocal arrangements for the design of curriculum content and standards, the exchange of information, students and teachers (Beazley, 1992, p. 1).

International form, on the other hand, refers to a program of study involving work experiences or a study visit abroad, or involves teaching taking place in a foreign language. Within this category it is assumed that the content is international as well. Apart from the impact of international curricula with a given course, the student must handle the language and culture of a different country.

An international dimension will typically include one or both of these aspects of content and/or form. Mestenhauser (1997), however, argues that it is precisely this kind of pluralisation of curriculum perspectives which makes the distinction between formal (content) and operational variables (method of instruction) difficult to sustain.

The OECD's agenda of internationalisation is not limited to OECD member countries. Rather, the OECD has actively sought to propagate its thinking to educational developments in other parts of the world. Indeed, it would be odd if it did not, since internationalisation implies developing scholarly ties across the globe. Moreover, student mobility is predominantly one-directional — from South to North — so OECD countries, for example, Australia and Britain, have sought to understand higher education in market terms. International education has increased OECD countries' revenue base from university education. Asia has been important within this commercial context. However, the OECD does not view Asia only in terms of new educational markets but also as part of a new global configuration of nations. The global flows of technologies, finance, images, and people documented by Appadurai (1996) have attained a volume and a value that they require greater collaboration between regional and economic groupings. In education international collaboration is part of these global trends. Recognition of this led IMHE/OECD to sponsor a conference in Melbourne at October 1996 on the *Internationalisation of Higher Education in the Asia–Pacific Region.*

Building on two previous conferences on this theme, one in Washington, D.C., in 1994 and the other in California in 1995, the aim of the Melbourne conference was to extend the work that IMHE was already doing into developing institutional strategies for the internationalisation of university education. It was felt that a dialogue ought to be established with the nations of the Asia–Pacific region, which included the homelands of many international students at European and North American universities. Australia was thought appropriate for the conference because of its regional links and its commitment to the policies of internationalisation of university education.

The opening speaker argued that Australia had to overcome a perception that its universities were interested more in education as a source of export revenue, and as a market commodity, than it was with the mutual benefits of international education within an emerging global economy. A view of internationalisation narrowly commercial and with insufficient recognition of student needs and insensitive to the benefits of international education did not serve the nation well, particularly as Australia was set on becoming a more open, internationally competitive, and globalised economy. Internationalisation, he noted, should highlight the increasing interdependence among nations, which implied an education system that facilitated partnerships and the exchange of ideas, skills, staff, and students, within a diverse and sophisticated global environment.

This analysis had much in common with the ideology of globalisation put forward in position papers by representative of Hong Kong, Indonesia, Thailand, Malaysia, Singapore, the Philippines, Japan, and South Korea. Most speakers assumed that the globalisation of the economy implied a particular way of managing university education. Internationalisation of education was thus interpreted both as an expression of, and a response to, the broader processes of globalisation. No attempt was made, however, to analyse these processes: they were taken as self-evident.

Yet, an analysis of global imperatives set the background against which Knight and de Wit developed a framework for comparative assessment of the range of strategies employed nationally and institutionally to internationalise university education. Their framework — a form of benchmarking — was also intended to provide a forum to discuss current issues and future trends (Knight and de Wit, 1995). In as much as Knight and de Wit's (1995) evaluative framework highlighted the need for all universities to adopt a similar pattern of organisational development, it assumed a "single idealisation of appropriate organisational behaviour", whereby "organisations must have the capacity to make a flexible response to uncertain market conditions caused by commodity saturation". This means in part that organisations are now required to become more competitive and develop new "educational products" to stay "ahead of the pack" by creating new markets (Waters, 1995, p. 81). This vision also implies a particular form of university education, sensitive to the need to prepare graduates to enter global workplaces, graduates who are confident in their capacity to move across national boundaries and to relate to a diverse range of cultural practices and traditions.

It is no less evident, however, that the OECD's discursive interventions have woven together commercial and educational aspects of internationalisation. The view of international education grounded in a range of market strategies to increase the number of international students and to expand the financial base of universities has been spliced together with a view of a global university, characterised principally by its engagement with the processes of globalisation, by its international networks and by its internationalised curriculum. The OECD contributed to the maturation of a discourse of internationalising university education that has developed in recent years, together with greater recognition of how in a singular manner it spans the cultural, economic and interpersonal dimensions of global relations.

Today, the context in which higher education evolves has been re-shaped by globalisation of which much has been said and written in recent times. Some of it is grossly exaggerated. But a great deal seeks to understand the profound changes that are helping to integrate the world into one global system. Developments in information and communication technologies, for example, involve knowledge production and exchange that transcend traditional disciplinary and cultural boundaries. They have brought about a major shift towards international integration of products and markets. National institutions are still significant in the global environment, though they must now become engaged in global processes or face obsolescence.

International competition and technological change are associated with a workplace more integrated and more devolved, that requires higher levels of cognitive and communication skills. The post-Fordist vision of flatter organisational structures demands high participation, strong teams, multi-skilling and lifelong learning in order to stay competitive. The future of work is increasingly shaped by technology, the capacity of labour and change management. The contemporary context is also characterised by the changing global knowledge economy (Castells, 1996). Among other features this includes: an exponential increase in the amount of knowledge; an acceleration in the rapid movement of knowledge globally; a growth in the centres of knowledge creation; a huge development in knowledge-mediated industries and services; changes in the access to, and control over, knowledge; and the emergence of new ways of thinking about the links between knowledge and innovation. The traditional links between knowledge and culture are also changing, with a greater recognition that knowledge creation and use is mediated by cultures. The changing nature of the global knowledge economy involves intricate and reflexive global–local relationships. It suggests that the nature of knowledge use and innovation demands a simultaneous engagement with local factors as well as global processes. This is so because in cultural terms the local is now being re-shaped globally, and because the idea of the global is meaningless without its local references and touchstones.

These remarks highlight the importance of looking at globalisation through the lens of the changing nature of the social relations it engenders. In this new setting, the boundaries of nationhood, geography, ethnicity, class and gender become fluid and shifting. The changes experienced come partly from increasing exposure to cultural diversity through the influences of international news and media, information and communication technologies, consumer products, as well as greater mobility. These increases in cultural globalisation are experienced as pressure tending towards simultaneous heterogeneity and homogeneity, a resurgence of local cultural identities, as well as the coalescence of globalised cultural practices. The global context is defined by a discourse that brings forth the cultural aspects of economic relations, and the need to develop products that reflect local needs, values and traditions.

In terms of these considerations about globalisation, the OECD's interventions into debates about the internationalisation of education are constructive. However, these interventions become highly problematic when viewed in the light of its predisposition to interpret the sociology of globalisation in a one-dimensional manner. The OECD has suggested that the internationalisation of education may be viewed as a response to the processes of globalisation. However, the relationship between what might be held as the global context and educational goals is not simple. What is represented as "the context" is never self-evident. It always requires interpretation (cf. Seddon, 1994). Descriptions of global processes are highly controversial, as are the constructs used to explain them, to respond to them, to react to them, or indeed to use them for comparative competitive advantage.

In terms of the internationalisation of education, the OECD fails to take into account a curriculum that seeks to provide students with skills of inquiry and analysis rather than a set of facts about the "new realities" of globalisation. In the context of a fast-changing knowledge economy, students need to develop questioning skills to enable them to identify the sources of knowledge, assess the basis of its legitimacy, examine its local relevance and significance, determine its applications, and speculate about how it might be challenged. The ability to think reflexively and critically about knowledge creation and application demands a form of global imagination, in short, the capacity to determine how knowledge is globally linked, no matter how locally specific its applications.

The internationalisation of university education involves a melange of global processes concerning conceptions of knowledge, economic exchange, the changing nature of work and labour requirements and cultural diversity. Internationalisation is relevant to all facets of university life, including teaching and learning, research and development and institutional management. Since it is so, it demands an holistic approach to change. It does not affect international students only, but is also relevant to the experiences of all students and all staff. With increasing global flows in communication and movement, we are now all influenced by globalisation. Globalisation can be represented as the transformation in which various practices are increasingly geared to operating in international surroundings, under international market conditions and with an international professional orientation. If this is so, then internationalised curricula involve the development of new skills, attitudes and knowledge among students and staff alike. They require the creation of new learning practices, spaces, ethos and cultures. This cannot be done by a university edict, but through the creative utilisation of the imagination of all those who make up a university. This imagination itself needs to be globalised in ways that are both self-reflexive and critical (cf. Appadurai, 1996). Internationalisation of the curriculum should therefore be seen as a dynamic process that gives staff and students the opportunity to own the processes of their own learning and knowledge production.

This view should greatly extend and re-shape our current understanding of the idea of internationalising of the university curricula. Internationalisation of university curricula should not only refer to those curricular activities that are designed for international students, or courses offered off-shore, or curricula leading to joint or double degrees with overseas partner institutions. Nor should it be restricted to interdisciplinary studies of particular regions of the world such as Asian or European Studies, to curricula that lead to qualifications which are internationally recognised, to the addition of a comparative dimension to traditional content or indeed to programs designed to enhance international students' capacity to survive mainstream university curricula.

The idea of internationalisation of curricula should be much more radical than this, and encompass the integration of a global perspective in curriculum design, development and evaluation. Curriculum content should not arise out of a singular cultural base. It should engage critically with the global plurality of the many sources of knowledge. It should respond to the needs of the local community, and give students knowledge and skills that assist their global engagement. It should encourage students to explore how knowledge is now produced, distributed and utilised globally. It should develop an understanding of the global dimension to economic, political and cultural exchange. In short, it should advance global understandings and global imagination.

Internationalisation of curriculum should not focus on content alone. It should include pedagogy and cross-cultural understanding. With demographic changes in our classrooms, how to cater for and take advantage of individual and cultural differences in learning ought to be crucial in the development of effective university pedagogies. New communication technologies have created new seats of learning which may link students to the global networks of information and ideas.

Through internationalisation, cultural diversity becomes a permanent feature of university life and a national strength. Diversity is an essential characteristic of a dynamic and creative society, able to engage effectively with global forces and to meet the challenges of the new century. Internationalisation of curricula ought to include values of openness, tolerance and cosmopolitanism. It ought to demand culturally inclusive behaviour, to ensure that cultural differences are heard and explored, that curriculum reflects the determination to learn from other cultures and that a wide variety of factors affect cultural change.

Finally, the internationalisation of the curriculum requires both students and staff to become more self-reflexive about what they learn and teach and how. It demands new practices of assessment and evaluation that are culturally sensitive and inclusive. Such assessment practices should reward innovation and critical engagement. If universities are to prepare students for a world of the ever-changing economies of global knowledge and social relations, then the goal of 'professional and vocational education' should be defined in terms of the global nature of work and economic and cultural exchanges, and be based on the premise that these elements are subject to continuous change. Preparing students to see change as positive and to manage it effectively in a global context should be a central aim of internationalised university curricula.

Conclusion

Some of the ways in which the OECD exerts influence over the educational policy making processes of nation-states have been examined and its discursive interventions as an international mediator of policy knowledge brought to the fore. These tasks involved an examination of its 1998 Thematic Review *Redefining Tertiary Education* and the OECD's active promotion of internationalisation of university education. The discussion illustrated a number of facets about the way the OECD works and about the educational ideologies it has promoted over the past two decades. First, the OECD pools ideas from its member countries, but in a way largely pre-determined. Some leeway for disagreements exists, but it is limited; and contrary views can often be accommodated within its generalised prescriptions. This ensures that its over-riding ideological position is not disturbed. Second, the OECD has developed a range of sophisticated mechanisms for influencing the thinking of policy makers not only within its member countries but also beyond. Conferences play a major role; as do the reports that emerge from them. Trend reports and the mediation of a policy language and concepts are also important elements of influence. Third, the OECD's agenda for the internationalisation of university education is an amalgam of commercial, political and cultural concerns that reflect its distinctive view of globalisation, as does the Thematic Review. Fourth, its agenda for the internationalisation of university education reinforces its (largely) instrumentalist view of education, as means for preparing students for a global economy in which neo-liberal and corporate managerialist ideologies are dominant. The Thematic Review attempts to expound a more nuanced, and/also approach to conceptualising university education. Universal participation in tertiary education is deemed necessary to ensure rewarding careers for individuals, to overcome youth unemployment and to ensure the competitiveness of national economies within the global economy where knowledge has taken on greater significance.

Finally, the OECD espouses a particular view of globalisation that is itself not subjected to critical analysis. This is clear from the deconstruction of the Thematic Review and from analysing the OECD and the internationalisation of university education. Globalisation appears as a given reality, the nature of which educational policy makers are implored to understand and respond to. However, the OECD's ideological leanings are reinforced through its (unreflexive) understanding of globalisation, couched as it is in a discourse of neo-liberalism and corporate managerialism. In establishing such a discursive framing for university policy, the OECD functions as a policy actor anticipating further global educational policy convergence.

8

The OECD and Educational Politics in a Changing World

"The role of the OECD in the new world, the 'global village', is growing. The collapse of communist regimes, the transformation of Asia, the move to market economies, the freer flow of goods and services and greater diffusion of capital and technology have all been part of the potent recipe producing this more open, 'smaller', globalised world."

(OECD, 1998a, p. 5)

"[Education] will play its crucial role all the better by awakening and sharpening critical intelligence and by allowing individuals to move beyond fear, introversion and ethnocentrism, which are the secret temptations of all societies. It is only at this point that civil society takes on its full significance, for civil society is not possible where citizens are not free and responsible, and where there is no education in the true sense of the word."

(OECD, 1998d, p. 117)

This study has set out to examine the ways in which the OECD has articulated, responded to, and been affected by globalisation processes and pressures with a specific focus on its education policy agendas. It has also delved into the implications of globalisation for national educational policy production. This chapter brings together some of the unfinished business of previous themes by discussing some key dilemmas that globalisation posed for the OECD and more generally for educational policy processes and priorities.

These dilemmas are essentially political in nature. They raise questions about the loci of educational policy making and the values which should sustain policy decisions. They point towards a bigger problematic of educational politics for/in a world where the nation-state has lost some of its policy salience before the new political groupings which operate above and below the nation-state. The political, cultural and economic configurations of globalisation do not deny a space for national policy making — far from it — but they have certainly altered the setting within which national governments establish their policy priorities. They have added new layers to the processes of policy production. Opening up some of the issues involved will serve both to recapitulate arguments developed previously and to speculate upon possible future directions for education policy.

Dilemmas for the OECD

The OECD has been both a globalising agent as well as being subject to pressures from globalisation As a globalising agent, it has been a vigorous exponent of economic globalisation, framing what has often been referred to as a new policy consensus in education, that is, the view of education as a key component for countries' and individuals' competitive advantage in the global economy, and of the necessity of finding more efficient and effective mechanisms of educational governance. Within the Organisation's broader ideological task of promoting market liberalism, the social dimensions of societal organisation have not been neglected but they have been incorporated into a paradigm in which liberal economics has achieved metapolicy status. Within the OECD, education and economic policy have become elided, despite the Organisation's parallel concerns with the social purposes of education, a situation which reflects in Stephen Ball's phrase: "a concomitance, if not a correspondence, ... between the logic of globalization — as a world free-trading system — and the new terrain of thinking about social policy" (Ball, 1998, p. 124).

However, the OECD is also the subject of the effects of globalisation and as such its established niche has been destabilised. The OECD grew out of a particular moment — the 'new international order' following the Second World War — and was heavily involved in rebuilding Europe during the Cold War. It was part of the hegemonic project of the United States to create liberal democratic societies, then largely framed in Keynesian terms. With the Cold War ended, the situation is very different as the OECD attempts to position itself within a 'new global order'. As it does so, it is squeezed by new regional organisations, by the EU but also NAFTA and APEC, and new, possibly more competitive, relationships with other international organisations. Moreover, it is operating with a changing and expanded membership which calls into question its traditional constituency and thus its relative homogeneity, exclusivity and consensual mode of operation. The dose of austerity medicine to the OECD's own operations — and the subsequent downsizing, budget-paring and modern management emphasis on "eliminating waste" (Sullivan, 1997, p. 111) — has put pressure on the liberal think-tank. With fewer staff having to meet increased demands, the Organisation has had to look to other sources of income, which has made its function as a think-tank more of a luxury. In this new climate, this analysis suggests that retaining a *distinctive* role places the Organisation before some strategic dilemmas in terms of how it is to approach its educational work and how it frames its educational priorities.

Neo-liberal contradictions

The ideological struggle between the contending economic systems of capitalism and communism has collapsed, leaving as yet no superpower to rival the US. Paradoxically, however, the contradictions of market liberalism are beginning to surface, generating new misgivings from both Left and Right as to social stability and social exclusion. These preoccupations now pervade all the major international organisations as well as individual countries. Their importance is not lost on the Secretary General of the OECD who argued that "necessary restructuring and efficiencies must be applied with a sense for their immediate social consequences, and with an eye toward mobilising public support for globalisation. Joe Public must feel himself part of the process" (Johnson, cited in Sullivan, 1997, p. 109). Social exclusion transcends the North–South divide. Increasing inequalities in wealth are found within as well as between countries, with poverty now marbled across societies. This development is connected to the ways in which capital circulates through global cities or regions, benefitting only some citizens while excluding others. In many ways the agenda of social exclusion is framed residually in terms of trying to address the problems of economic and social polarities without modifying the fundamental tenets of neo-liberalism. In education, the discourse of exclusion may serve as a new framing. It is a contested terrain, however, which stretches from a broad focus on redistributive policies in a global context to a more narrow concern with 'employability' and social cohesion.

In this setting, how might the educational work of the OECD face up to the contradictions in which the Organisation, as itself a globalising agent, is intricately implicated?

The hollowed out logic of performativity

An increasingly prominent part of the OECD's educational work has been given over to promoting what we have called a 'culture of performativity', a central component in new forms of governance and modes of accountability. In its current form, performance has to a significant degree been reduced to quantitative measurements of student or system outcomes or applied, increasingly, to comparisons of national systems. While the will to measure can in part be traced to an ontological sense of 'improvement' embedded in Western technical rationality, the current framing of performance-as-accountability brings with it the dangers of confusing means with ends, style with substance — in other words, of a 'hollowing-out' of educational purposes to sets of numbers or ticks on a checklist. More fundamentally and reflecting the epistemological relativism of postmodernism, in its exaltation of style — performance management — over substance, performativity serves to legitimate a cynicism vis à vis claims to universal truth in favour of a smorgasbord of values. Thus the capacity of education systems to distinguish between superficiality (as expressed for example in trite versions of 'internationalisation' of the curriculum, in airbrushed institutional profiles of performance, or in the monies poured into elaborate websites and other devices of institutional branding and marketing) as against the achievement of fundamentally important goals such as equality of opportunity and outcomes, is called into question.

How, then, can the OECD address the contradictions of performativity? How can it argue for a progressive reading of 'improvement' in the face of a logic which epistemologically is impotent to distinguish superficial accountability and the easily quantifiable from compliance with substantially important goals?

The liberal think-tank trap

The OECD's original charter promoted market liberalism and pluralist democracy. It has served as a source of strength, but also of discursive constraint. Here is a paradox indeed. On the one hand, the Organisation's reputation and indeed its standing rest on its analytical capacities. Its cachet is linked to its elevation above the machinations of national politics. At the same time, its ideological mission renders it unable to envisage genuinely alternative paradigms. Such a constraint is particularly telling in light of the Organisation's changing constituency, and the changing problems the Organisation has to address — for instance, the presence of enclaves of Third World conditions which now exist in First World countries (and vice versa). Such changes impose a shift in the Organisation's *raison d'être*, from think-tank role to problem-solving role. The latter is embedded more in the immediate problems which member countries wish to solve and to which the OECD seems more willing to provide solutions, as in its consultancy outreach work with non-member countries.

To provide such solutions, however, may require a radically different frame of reference. How, then, can the OECD in its educational work envision alternatives? Can the think-tank think new truths?

Arguably, these dilemmas point to the need for a new frame of reference for education — one which, given the increasing interlinkages between education policy and other policy domains, in turn connects to broader struggles to shape the processes and outcomes of globalisation.

The Politics of Globalisation

The discussion here draws in part on developments in social policy theory. In particular, it draws on arguments that the globalisation of social issues such as peace, the environment, or social welfare together with the backlash against economic globalisation, may provide the seedbed for a frame of reference alternative to hegemonic market liberalism which dominated the policy landscape over the last part of the 20th century.

Two interrelated issues, of social cohesion and political accountability, are of special relevance. A small example — the defeat of Australia's Republican referendum in 1999 — may serve to reveal what is at stake. In that referendum, the successful 'No' campaign pivoted around, not the retention of a constitutional monarchy, but the slogan 'Say no to the *politicians'* republic' (original emphasis — which referred to the fact that the republican model represented a procedure for appointing a head of state by parliament rather than direct election). The defeat was summarised in *The Australian*, the Commonwealth's only national daily newspaper, in these terms:

> "The defeat of the republic exposes Australia as two different societies — a confident, educated, city-based middle class and a pessimistic, urban and rural battler constituency hostile to the 1990s change agenda. This schism is not just an insuperable obstacle to a republic. It is far more serious — a threat to a cohesive and successful Australia as it tries to adapt to the globalised economy of the new millennium."
>
> (Paul Kelly, *The Australian*, 8/11/99, p. 1)

Though a somewhat simplistic account, the evocation of the 'two societies' syndrome has been widely recognised. It is not, of course, confined to Australia as Martin and Schumann (1997) indicate in their popularisation of the idea of the 20/80 society. While Levitas (1996) pointed out that such a depiction neglects the considerable degree of inequality within the 80%, the fact remains that economic globalisation is producing great polarities of wealth and poverty everywhere, with their attendant social consequences including social instability, backlash political movements and electoral volatility. Backlash movements have been fed, additionally, by the higher levels of cultural diversity flowing from the symbiosis of cultural and economic globalisation. The politics of globalisation, therefore, involve addressing "the challenge of valuing diversity" (OECD, 1996d, p. 30) along with profound social and economic inequalities and associated problems of alienation from political activities.

Globalisation and social exclusion/cohesion

The devastating social fallout of globalisation is widely acknowledged. What remains in dispute is how this situation should be analysed and addressed. For the advocates of globalisation, dealing with its consequences is a matter of 'managing change' — more explicitly, of managing problems of economic marginalisation, social exclusion and heightened levels of cultural differences within societies in order to enhance social cohesion. Such a programme frames these problems in terms of the existing paradigm of economic liberalism and growth. This is the OECD position, warning against throwing out the economic baby with the social bathwater. From this it follows that the OECD's prime goal is to maintain the momentum for fiscal and trade liberalisation in the face of 'backward' protectionist and nationalist stances and to find ways of distributing better the benefits of globalisation — the trickle down theory of economic growth. Within such a framework, education becomes one of the strategic tools for the management of change inasmuch as exclusion is cast in terms of a failure to engage with the global economy (in turn interpreted largely as a matter of lack of appropriate skills or disposition). Social tensions are held to the product of a lack of tolerance which can be overcome through policies and programs of multiculturalism and internationalisation.

For the critics, this is a residual framing for social cohesion, particularly so in addressing the core problems of economic marginalisation and social exclusion. Levitas (1998), for example, argues that much of the interest in social cohesion in the EU stems from an essentially functionalist Durkheimian view about the solidifying effects upon society of the division of labour. Within this perspective, Levitas argues, social exclusion is defined in terms of unemployment. The solution — integration — is seen in terms of integration into paid work, a framing which excludes, for example, many women. The problem with the discourse of social exclusion, according to Levitas, is that the emphasis on employability tends to obfuscate an analysis of poverty as deriving from fundamentally unequal relationships of capitalism. In consequence, she argues, integration into work becomes an end in itself without any interrogation of the conditions under which such integration may occur or, more fundamentally, the deep-seated structural inequalities of (global) capitalism "driven by profit and based upon exploitation" (p. 18).

An alternative to the discourse of social inclusion deriving from a structural and redistributive approach, Levitas suggests, takes as its starting point poverty and attendant problems of exclusion from civil society and citizenship rights. This is closer to the conceptualisation that underpins the discussion by Bessis (1995) of social exclusion and cohesion. It was based on the conclusions of a symposium sponsored by a number of international organisations including UNESCO, the ILO, the World Health Organisation and the Commission of the European Union. Bessis identified three dimensions of exclusion: economically, in terms of lack of employment; socially, in terms of loss of social status (deriving from unemployment and loss of income); and politically, in terms of deprivation of political and human rights. In part the discussion is framed in economic terms. The solution is not however seen, solely, in terms of integration into the labour market, but rather in terms of confronting what she describes as "the dictatorship of the economy" (p. 26) and "the reign of the *pensée unique* — a single acceptable way of viewing things — in the area of economics" (p. 13). For Bessis it is strategically important to "put an end to the hegemonic status of dominant economic thinking ... and give primary to social policy" (p. 35). From this perspective, the goals of educational policy become framed by a commitment to building social capital as well as to human capital investment.

The concept of social capital has received a good deal of attention in social and welfare policy as a way of setting down an alternative starting point for policy goals. Thomson (1999) for example suggested that the interest in social capital stems from three impulses: from a response to the dominant individualism underpinning the development of human capital for purposes of national competitiveness; from a recognition that economic success requires a certain level of social cohesion, stability and trust; and from a growing recognition that many people are dissociating economic success from a sense of well-being. She defines social capital as "the outcome of social processes which link people together in groups and build communities". She develops her argument further:

> "A society with high social capital may be characterised as having a rich web of social and civic networks and a low degree of social division. Social capital can be created by public policies that facilitate, build on and institutionalise local social networks, but it is destroyed by public policies that disrupt, ignore and destroy the local social fabric. Local social networks are created by the actions of local institutions and local residents. When public policy values local social networks and democratic processes and builds them into programmes and decision making, then they become part of the wider social fabric and can be categorised as social capital."

> (Thomson, 1999, p. 2)

In a similar vein, Hadenius and Uggla (1998) define the concept of social capital as referring "to various forms of linkages or bonds that hold people together. In other words, social capital is a collective entity: it is the 'cement' that makes individuals part of a group or a community" (pp. 43–44). These bonds, they suggest, have two dimensions: affective and institutional. Affectively, social capital serves to promote organisational capacity by which people learn to coordinate activities in order to bring pressure to bear on other groups or on the state. Such experiences in turn help to promote tolerance of different viewpoints in a pluralist society and, further, the skills acquired may be generalised into broader political action. Following a view of De Tocqueville (1835) of democratic politics they suggest: "this nurturing effect of involvement in civil society is the keystone of democratic politics: people develop practices in the private sphere which enable them to play the roles of political citizenship" (p. 44). They cite the example of the Bangladeshi Grameen Bank which, in addition to providing credit to the poorest sections of the rural community, by its active promotion of group formation contributed to nearly doubling the number of women who vote (p. 44).

Social capital formation draws strength from participation in civil society — that is, in community associations operating independently of state authorities. The central point in this discussion is not only a reprioritising of policy values, but also the revival of civil — or civic — life and local/community politics which, critics argue, have been significantly undermined by the eroding effects of individualistic neo-liberalism. Preoccupation with social capital formation and social cohesion returns us to the issue of the role of state and its function, which according to Putman (1993) is to provide the structures of institutional support that enable a healthy civil society to flourish. More broadly, what is at issue are the appropriate relationships to be forged between the principal actors on the stage of globalisation: the market, civil society and the state.

One attempt to develop new relationships between these actors can be found in so-called Third Way politics, associated with some European social democratic movements (especially Germany) and the British Labour Government under the influence of Anthony Giddens (1998). Giddens describes the Third Way as a renewal of social democracy which "has to be left of centre, because social justice and emancipatory politics remain at its core" (p. 145). The core concern with social justice is defined largely in terms of social inclusion and exclusion. Giddens argues that new ways of social organisation and welfare provision have to be found to offset the socially damaging consequences of neo-liberalism and the potentially anomic effects of globalisation. The Third Way charts a path between old style welfarism with its topdown, bureaucratically organised distribution of benefits and the harshness of neo-liberal deregulation. It "looks for a new relationship between the individual and the community, a redefinition of rights and obligations" (p. 65). Giddens' path echoes to the sound of ecclesiastical zeal:

> "Positive welfare would replace each of Beveridge's negatives with a positive: in place of Want, autonomy; not Disease but active health; instead of Ignorance, education, as a continuing part of life; rather than Squalor, well being; and in place of Idleness, initiative."
>
> (p. 128)

Giddens' vision is in many regards a thrusting, driving vision of the globalisation roller coaster in which the role of the state is to provide not welfare, but the platform from which communities may bond and individuals may flourish. Thus the welfare state becomes the social investment state in which investment in human capital is seen as a core activity. Welfare becomes "positive welfare to which individuals and other agencies besides governments contribute" (p. 117). The "risk society" (Beck, 1992) is elevated to a new good; even old age is presented as a "positive risk" requiring more creative (self help) approaches than old style (welfarist) pensions. Poverty programmes are replaced with community-building programmes emphasising "support networks, self help and the cultivation of social capital as means to generate economic renewal in low income neighbourhoods" (p. 110).

Those core values underpinning Giddens' Third Way include equality (defined as social inclusion), protection of the vulnerable, no rights without responsibilities, no authority without democracy and a cosmopolitan pluralism. In practice, what has been most remarked and talked about, particularly in the Anglo-Saxon world, is the rhetoric of "no rights without responsibilities" — of "mutual responsibilities" — prompting criticisms of such politics as simply individualistic liberalism reworked. As such, the Third Way is regarded sceptically in many parts of Europe by those seeking a more robust formulation of social democracy. However, Third Way politics should be seen as part of a broader struggle to find new frames of reference and new modes of governance able to address issues of social inequality and marginalisation, cultural diversity and political alienation that flow from the processes of globalisation.

Globalisation and political accountability

One of the core debates surrounding globalisation has to do with its democratising potential. On this, there is an ideological divide. Enthusiasts of globalisation see power as being democratised through the new cultural and political formations resulting from the flows of globalisation — the internet, new diasporic communities, social movements whose influence has been extended through the new communications technologies, reinvigorated local politics accompanying the devolution of centralised systems and polities, and so forth. Friedman (1999), for example, in his popular account of globalisation in *The Lexus and the Olive Tree*, perceived the benefits of globalisation stemming from the three-pronged democratisation of communications/technology, finance, and information (though he acknowledges a hefty problem in finding ways to distribute the benefits). For the OECD, the assumption of a nexus between economic and political freedom has been embedded in its charter from the outset and forms a key plank of its strategy of globalisation.

For critics of globalisation, globalisation produces a democratic deficit. Markoff (1998), for example, argued that within the polity of the nation-state, the "institutions of modern democracy and the modern politics of the street developed in tandem" (p. 5). State institutions of power, he argued, required legitimation from below, and forced an engagement between the political elites and the people. By contrast, he went on to suggest, transnational structures associated with globalisation make this process of engagement — never very complete — more difficult. For Markoff, the problem remains of finding ways to democratise power such that it is commensurate with the globalisation of capital.

> "The emerging structures of decision-making ... do not have such features [as separation of powers; democratic legitimation such as elections] and much of their activity is even hidden. The inner processes of the World Bank and IMF ... are hardly publicized and positions taken by many national representatives are not even made publicly available. Rather than legitimacy, it is invisibility that is sought. How such power might be democratized is the challenge of the twenty-first century."
>
> (p. 26)

Neave (1997, p. 22) similarly refers to the weakened link between economic development and democratic accountability, and asks whether globalisation implies:

> " ... an even greater uncoupling between the politics of economic development, conducted by international agencies and urged on by cross national enterprises, and the politics of democratic accountability and participation which are conducted still within the setting of the individual nation or State."

Although sharing the pessimism about economic globalisation, our own view is that such depictions oversimplify the multi-dimensionality and internal contradictions of globalisation processes. There is little doubt that economic globalisation is exacerbating social inequalities. However, cultural and political globalisation processes work more ambiguously. For example, globalised technologies reflect in many ways the cultural and linguistic hegemony of the United States, but simultaneously offer possibilities for more democratic, low cost access to education and other forms of social participation (Kishun, 1998). Or to take another example: the greater cultural diversity resulting from enhanced flows of people exacerbates social tensions. But cultural flows also enable the flourishing of new diasporic public spheres and politics (and for the mobile elites, greater work opportunities in the global marketplace). Moreover, the polycentric nature of political globalisation, facilitated by the new technologies, opens up new possibilities for, and new modes of, political organisation. We would take issue with the pessimists, not so much over the facts of the democratic deficit as with the implications they draw about the political landscape.

One problem with the critics' formulations is that the dichotomies between (the good, democratic) nation-state and (evil) international agencies, accountable only to the interests of the winners in globalisation are overdrawn. There is nothing inherently egalitarian or democratic about the nation-state. Indeed, one of the arguments of the proponents of globalisation is that the bringing in of command economies under the economic liberal umbrella helped to engender more democratic domestic politics. Likewise, international organisations are by no means ideologically aligned or straightforwardly unthinking proponents of free trade and economic liberalism and growth. They are not necessarily captive to the big end of business. International organisations also play key roles in the promotion of egalitarian and humanistic agendas, for example in the areas of peace and human rights (the latter now one of the eligibility criteria for promotion to the OECD 'club'), the environment, labour market and social policy. Even the OECD is not ideologically homogeneous.

In this context, the concept of a polycentric political sphere takes on some salience. Deacon et al. (1997), for example, in their discussion of the "socialization of global politics" (p. 3) point to "a new paradigm of political science [in which] tiers of government in effect give way to spheres of influence of complementary and contending local, national, supranational and global political forums" (p. 6). Such a scenario begs the question as to the nature of the linkages between these spheres of influence and their effectiveness in the face of powerful global capital. But their general argument is that social policy is increasingly shaped by the politics of supranational agencies and non-government organisations which "are increasingly the locus of future ideological and political struggles" (p. 10). These agencies, they remark, work in contradictory directions. The struggle for better global and national social policies is a struggle of values and ideas within major international organisations (p. 201). The discussion by Walby (1997) of the place afforded to feminist politics around the social charter of the EU also points to a more nuanced analysis of the politics of international and supranational organisations. She argues, first, that the social charter enshrines better equal opportunities policies for women than is found in many of the constituent nations of the EU. Secondly, whether these policies will be harmonised upwards or downwards remains an open question. Thus, she suggests, "the position of women and the representation of gendered interests within the political structures of international or supra-national bodies as compared to the national is crucial to future developments in the EU" (p. 211). The continuing under-representation of women within the OECD and other international organisations highlights the need for such gender issues to be addressed.

Deacon et al. (1997) go so far as to conclude: "It is no longer a matter of being 'in and against' the state but of being 'in and against' international organizations" (p. 6). Again the dichotomy is overdrawn. More precisely, international organisations should be viewed as analytically part of an extended polity which incorporates national, sub-national and supranational elements, and requiring transparency in the ways these elements work and link together. This view is echoed in Bessis' comment that "to reorganize democratically a world that is characterized by the tradition of a centralized nation-state, the dilution of responsibilities on the international level, and the silence of local actors, it seems necessary to reflect on new relationships between the local, national, international and global levels" (p. 43). Bessis opts for more participatory forms of democracy embracing, at one end, scope for community action and at the other:

> " ... new forms of political and economic regulation on the international level ... Reform of the international system must put an end to the monopoly of interstate organizations and allow for the emergence of democratically-oriented global organizations with powers of oversight and of putting forth proposals."
>
> (p. 46)

Along with such a reconfigured polity Bessis argues for a less Eurocentric set of priorities. The globalisation of social problems — that is, the relative interpenetration of the North and the South — she argues, has indeed facilitated global thinking. "But if one is not careful this global thinking could once again take on the thinking of the developed North." (p. 46) The discussion points towards the need to dissect further the idea of global policy communities. In this context, notions of "upper and lower circuits of globalization" (Torres, 1995) and "globalization from above and below" (Falk, 1993) are useful. The ideas refer to the distinction between, on the one hand, the powerful flows of capital and culture which appear to be having a homogenising and determining impact on nations from above, and on the other hand, an emergent democratic grass roots politics based on notions of global communities and proactive citizenship.

Globalisation from above, in Falk's view, reflects:

> " ... the collaboration between leading states and the main agents of capital formation. This type of globalization disseminates a consumerist ethos and draws

into its domain transnational business and political elites. It is the New World Order, whether depicted as a geopolitical project of the US government or as a technological and marketing project of large-scale capital, epitomised by Disney theme parks and franchised capitalism (McDonalds, Hilton, Hertz . . .)."

(Falk, 1993, p. 39)

By contrast, the politics of "globalization from below":

" . . . consists of an array of transnational social forces animated by environmental concerns, human rights, hostility to patriarchy, and a vision of a human community based on the unity of diverse cultures seeking an end to poverty, oppression, humiliation, and collective violence."

(Falk, 1993, p. 39)

From a slightly different perspective, Deacon et al. (1997) also point to differing currents in globalisation by drawing a line between global social reformists and laissez-faire economists. They argue that a number of factors set the context for a new discursive framework which could "strengthen the hand of global social reformists against the hand of global laissez-faire economists" (p. 5). These factors include post Cold War geopolitics which creates new connections between North and South, East and West; the 'threat' of global economic migration (analogous to the impetus to national reform movements from the destabilising effects of the poor in earlier regimes of capitalism), and the influence of new epistemic communities within and across international organisations. As illustrations they cite the recruitment by the World Bank of new human resource specialists rooted in European social democratic traditions, or the community of defenders of labour and social standards gathered in and around the ILO and the Council of Europe.

Various possibilities for shaping the forces of globalisation in more democratic ways and from new reference points have been discussed. What are the implications that arise for the politics of educational globalisation — and more especially, for making education a vehicle for a more democratic expression of globalisation?

Educational Politics for a Changing World

We return here first to the idea of policy convergence across OECD countries around matters of educational governance and purposes. The discussion will focus primarily, though not solely, on the tertiary education sector, that is higher education and postcompulsory education and training.

As we have seen, elements in the new forms of educational governance include: (1) restructured bureaucracies; (2) new accountability mechanisms and modes of performance management such as quality assurance and profiling; (3) the creation of competitive markets and 'new partnerships' in education accompanied by reduced government funding and encouragement of other forms of income supplementation, for example joint ventures, various forms of user pays systems for domestic and overseas students; (4) a blurring of the traditional divide between universities and other post-compulsory education institutions in expanded systems of tertiary education.

New forms of governance both reflect and help to shape changing conceptualisations of the purposes of higher education. Mass tertiary education is now viewed primarily as a lever of economic growth with both individual and societal benefits. To that extent it is represented as serving social as well as economic purposes. Increasing levels of privatisation and commodification of tertiary education have marginalised non-commercial areas of inquiry and research, particularly in the humanities and social sciences.

The dominant policy paradigm depicted here is highly problematic — both in its own terms and in terms of broader goals of social justice.

Governance issues

First, the proliferation of accountability mechanisms associated with performance management has served to increase rather than decrease bureaucratisation, while adding new layers of performance demands. More fundamentally, performance management (or impression management) is to a very large extent hollow — that is, time consuming and feeding the will to measure rather than the will to know, to teach or to learn. While the rhetoric of new public management favours flatter structures and reduced hierarchies, corporate management in higher education has overseen the creation of new layers of (well paid) administrators — human resource managers, managers of technology, pro-vice-chancellors (academic, research, international, technology, wigs and gowns), executive deans and so forth. At the same time, teachers/academics and support staff struggle with longer hours, larger numbers of students, ever-reducing working conditions and declining salary relativities. New managerialism, which resonates more with the world of business and commerce than education, has contributed to delegitimating activities and values once associated with higher education — intellectual inquiry, social critique, curiosity-driven research, academic freedom and tenure.

Secondly, the rhetoric of new partnerships masks the extent to which education has been privatised, commercialised and further commodified. The formation of tertiary education markets is creating highly differentiated systems, not so much in terms of different 'missions' as in terms of wealth. Put crudely, multi-tiered systems are developing. The ancient divide between elite universities and other tertiary education institutions — polytechnics, colleges of advanced or further education, business colleges and so forth — persists though slightly reshuffled. New players are entering the field: virtual universities; corporate universities such as Motorola or McDonald's University; and new conglomerates such as that formed by the Cambridge–MIT merger, or the formation of Universitas 21. Many of these new species are likely to survive well in the Darwinian jungle of tertiary education, though possibly at the expense of smaller regional institutions. Finally, institutions are being internally segmented, along a teaching-research/consultancy divide, a situation exacerbated by cross-institutional alliances, be they whole research centres or other aggregates of academics and researchers.

Such alliances differ significantly from the traditional idea of the borderless community of scholars precisely because they occur in a highly competitive and commercialised context. (Of course, elements of intense competition have always been present in the scholarly community's pursuit of knowledge, particularly in the fields of medicine and science.) Yet these processes of dedifferentiation are generating heightened (not new) divisions between knowledge reproduction (teaching) and knowledge production (research) and, within the latter, between so-called pure research and scholarship (now very poorly supported) and industry-supported research (the new partners). Paradoxically, the fragmentation of higher education systems and governance goes hand in hand with a form of academic concentration arising from the interconnections between the educational policy elites whose globe-trotting brings them together over and again in conferences, high level meetings, and negotiations — a peregrination which raises the question to whom and to which institutions these travellers may indeed be accountable. However, what is really being argued here is that, beneath this multi-layering and seeming fragmentation, a dual system of tertiary education is being created: one well-endowed which serves to credential the global mobile elites; the other serving, residually, to credential the mass of students, now funnelled into vocational degrees and diplomas. Against this backdrop, although 'user pays' education is represented as a personal investment in future employability, this policy framing raises significant questions of equity, of the purposes of education as a public good and of finding ways to create genuinely diversified systems of tertiary education.

Thirdly, educational markets in tertiary education represent a chaotic mix of anarchy and regulation. Governments still seek to steer education. Thus, institutions remain bureaucratically tied up in performance demands. Buccaneering vice-chancellors/CEOs, particularly those at the head of institutions thriving in the new competitive playing fields, see themselves as "hogtied by government regulation" (*The Australian*, November 6–7, 1999, p. 38) They seek to free themselves from government restrictions by privatising as much as possible. Such anarchy is hardly conducive to an effective (rational), diverse and equitable system of provision and accountability. Further, dependence on private (hence uncertain) funding feeds into short-termism and destabilises an academic labour market increasingly composed of outsourced or contracted labour — in other words, into system chaos. When commercial logic impels universities to attempt to bypass irksome restrictions and when governments attempt to distance themselves from the responsibilities of funding, the question must surely be raised as to who is responsible for the provision of public education — supposedly the bedrock of both the learning society and of the knowledge economy.

Educational purposes

One of the major achievements of OECD inspired reforms in education and training which seems to have been taken up everywhere has been to forge a new discursive taken-for-granted. This is the performative or entrepreneurial self (in keeping with new human capital theory, applying to both individuals but also individual institutions or even systems). In other words, despite its problematic nature, entrepreneurialism is perceived as the solution to the vexed issues of social exclusion and the key to competitive advantage in the global marketplace. This point requires further elaboration.

Clearly, the so-called knowledge-based economy generates new dimensions of inequality. As Urry (1998) argued, globalisation is creating "inequalities of mobility as opposed to the social and place-based inequalities of the past. This may be the real importance of the so-called globalization of contemporary social life and its transformations of time and space" (p. 16). Such a development is significant. It suggests a reworking of the historic class divide, in which the fordist mental–manual division of labour is replaced by the technology/knowledge-based inequalities of a so-called postfordist work order. The role of education in producing worker-subjectivities for the new work order takes on particular significance — a theoretical point made by a number of scholars in relation to earlier phases in capitalist development. With the rise of industrial capitalism, Johnson (1976) argued that the bourgeois attack on working class cultural values (implicit, he believed, in the introduction of mass schooling) was a rational attempt to inculcate the values underpinning the relations of capitalist production. Bowles and Gintis (1976) similarly observed that: "the perpetuation of the class structure requires that the hierarchical division of labor be reproduced in the consciousness of its participants" (p. 147). They went on to argue that education systems served as a significant arena for the socialisation of future workers. Desirable attributes of the industrial worker included industriousness, punctuality and obedience. For the executive they were problem solving and initiative (Braverman, 1974). In the postfordist, knowledge-based economy, the distinctions between executive and worker have supposedly become blurred. Initiative, problem-solving, team-building, risk-taking and entrepreneurship becoming key *desiderata* for all workers. In Marginson's (1999) view: "The key figure in late modern systems of government is the self-regulating, choice-making, self-reliant individual" (p. 25), a figure nestling nicely into the "prevailing welfarist ethic of self-responsibility" (p. 28) but who, be it noted in passing, sits in paradoxical juxtaposition to the corporate team player.

In one sense, this is nothing new. The entrepreneur was the backbone of capitalism, and many years have gone by since Callahan (1962) examined the cult of efficiency in education. However, the entrepreneur of industrial capitalism existed despite or outside of formal education, and the cult of efficiency was largely a US phenomenon as befits the heartland of contemporary capitalism. What is new is the way in which the production of the entrepreneurial self (or school or university), linked to the corporate world's discursive canopy of clients, consumers, customers and competitors, is institutionalised as a goal of education systems across all countries. Institutionalised entrepreneurialism feeds into, and feeds off, competitive individualism — and by extension, of individual institutions or even individual conglomerations of institutions. The establishment of Melbourne University Private in conjunction with Universitas 21 serves as an apt illustration of this dynamic.

Universitas 21 is a global network of eighteen major, research-intensive universities from Australia, Asia, North America and Europe in association with a number of global corporations whose names, for reasons of commercial confidentiality, are not released. Melbourne University Private is a full fee-paying private extension of the public university, paying a royalty and licensing fee for the use of the words 'Melbourne' and 'University' and outsourcing its teaching to Melbourne University. The vice-chancellor's vision is to be a key player in a global market:

> " ... if you're a regional institution you are just not going to be in the game, because the best knowledge workers are internationally mobile and want globally portable degrees ... Our comparative advantage is being an Anglophone university in the East Asian hemisphere, culturally diverse, global in reach and absolutely uncompromising in quality. We'd hope our students can get jobs anywhere in the world in competition with the world's best graduates."

> (Alan Gilbert, *The Australian*, November 6–7, 1999, p. 38).

This example is presented at some length because it demonstrates the problematic nature of current directions in policy being undertaken in the name of globalisation. At one level, the manoeuverings of Melbourne University weigh lightly on the overall functioning of tertiary education in Australia. At another level, Melbourne's move to consolidate with other research-intensive, globally focussed universities as part of Universitas 21 illustrates the rise of new hierarchies in higher education. In itself, differentiation may be no bad thing — if set within some positive vision by governments of the multiple purposes of tertiary education. However, the divide now exists not as an expression of government policy so much as a manifestation of the anarchic logic of market forces reinforcing the 20/80 divide mentioned earlier. As Marginson (1999, p. 30) pointed out, globalisation of education as a commodity results in a binary split which on the one hand sees some universities engaged in producing "an emerging global elite ... while the large majority remain mired in the more impoverished and more regulated nation-specific institutions".

Whether the beneficiaries of the new entrepreneurialism will be any other than the mobile elites — or whether mobility will be significantly extended in the global economy — remains to be seen. At present, a good deal of attention is being paid to expanding educational opportunities for all — hence the moves to mass tertiary education, internationalisation of education, improving educational pathways, lifelong learning. But nothing in these developments suggests that greater occupational or social equality is being delivered. Rather, what seems to have developed could be described simplistically as a two tier — core and peripheral — workforce. Conventional wisdom holds that it is the unskilled (or the inappropriately skilled) who are relegated to the latter. But the periphery is also heavily populated with university students and graduates performing basically machine-minding tasks in the new growth industries — retailing, tourism, hospitality, financial services — thus further squeezing out those less qualified. "Credentials creep" rather than greater educational opportunity may underlie present developments.

Clearly there are problems with the policy directions embarked upon, even in their own terms. Arguably, even more fundamental problems exist in respect of social justice. The entrepreneurial vision for education is, we contend, ultimately empty. It connects poorly to the need to develop social as well as human capital. It connects weakly to strategies for promoting social justice. Education for/in democratic societies does not turn around matters of access and participation or the management of difference and diversity alone. It also enshrines the idea that education has enduring public value and stands for something beyond simple, economic and vocational utility. The core issues in educational politics involve challenging both a deterministic globalisation ideology and its imagining. They involve implementing alternative educational visions for a more democratic world order. As Thomson (1999) explained:

> " ... a reductionist emphasis on education's role in developing human capital has resulted in the atomisation of privatisation, where each student is a potential business centre, bearing a stock of knowledge and skill, updateable through life-long learning, and convertible to economic capital. At the same time, the individual is being constructed, through advertising, as a global consumer. Education is at a critical point in policy-making, needing to meet the challenge of re-building the social capital purposes, processes of, and outcomes from education, recognising that the individual is more than the sum of these roles, having civil and emotional responsibilities in relationships, families, communities and globally."
>
> (p. 8)

Certainly, some discursive levers are to hand which can be drawn upon to argue in support of the necessary conceptual shifts. Indeed, the polycentric nature of global politics provides forums — at local, national and international levels — for giving voice to those arguments. While educational politics at local and national levels are relatively well developed, international educational coalitions of interest are weak in comparison with, for example, the women's movement or movements for the environment, peace, and Indigenous issues. Granted, these movements often incorporate educational aspects, and an incipient international-level politics in education is to be found in the presence of international education unions and a number of educationally oriented NGOs.

Still an important and largely unfulfilled role remains nevertheless for higher education as an actor in educational politics. In this domain, the role of organic public intellectuals takes on a particular significance. One of the chief dangers contained in the globalisation of tertiary education in its current form is the systematic delegitimation of intellectual inquiry and social critique. In the current politics of derision, higher education occupies a highly ambivalent position. On the one hand, it is being discursively shaped as the crucible of employment. On the other hand, being (highly) educated is discursively linked with membership of the elites — the winners of globalisation, who are seen as rolling the barrel of self-interest. Such cynicism ought not entirely to be attributed to globalisation incorporated. The authority of (university) knowledge is also under more general threat from postmodernist epistemological Angst over universal knowledge claims. Yet higher education above all, and especially faced with the social consequences of globalisation, has a distinctive role to play in furthering debate, discussion and in the imagining of alternative visions.

Conclusion

This enquiry has sought to provide insight into the complexities of educational policy making in a context now often referred to as global. In part, it has been a case study of the ways in which globalisation processes and ideology have both shaped, and been shaped by, the OECD. The OECD is not a singular entity with a singular educational agenda. Rather, it exists in a complex relationship with its members being simultaneously policy instrument, forum and actor. In part, this study sought to show how and why the educational priorities of so many OECD countries have shifted and converged within the context of globalisation. In making this case, it has not been suggested that convergence equals homogeneity. On the contrary, in educational policy making — as with the broader politics of globalisation — the evidence points to a great deal of local/national variation in how convergent policy agendas are played out. Such variation stems both from active resistance to global trends and/or from the weight of national traditions, politics, culture and histories. The raging politics of nationhood in Europe, precisely because of the threat of 'Europe' and, beyond that, of globalisation, are a case in point.

As this chapter noted at the outset, two basic questions have been raised. They relate to the loci of educational policy making and the conceptualisation of educational purpose. To conclude on a more normative note, the nation-state still is, and ought to be, the prime site of education policy production, serving important national purposes of social justice — including economic prosperity for all citizens — and social cohesion. It is a view that also lies at the heart of the OECD's way of operating. Nevertheless, given the new cultural flows within and between nations, given the broadened polity of supranational and international bodies as well as national legislatures now involved in considering and mediating the affairs of nation-states, including educational directions, clearly nation building has to be a far more cosmopolitan enterprise than it was in the past. These trends, too, are reflected in the Organisation's policy activities.

The key issue is less the fact of policy convergence across nations, so much as the substance of what is subjected to that convergence. If convergence ushered in a more peaceful, environmentally responsible, better educated and better-fed world, few would complain. But this has not been the case. Rather, convergence has focussed around an ascendant neo-liberal paradigm of policy in which education has been largely (though not solely) framed as human capital investment and development. Such a paradigm serves to legitimate a set of educational values feeding off and feeding into the broader culture of rampant individualism and consumerism unleashed by the ideological victory of capitalism over communism. Educational purpose has, in large measure, been reduced to a student's calculus of job opportunities or to the state's calculus of maximum return on minimum input. Both echo the broader processes of the commodification of what was once regarded as 'the public sphere'.

Despite its broad-ranging work in education, the OECD has been a key player in sponsoring such a set of values, to which the Organisation's Secretary-General, Donald Johnson, gave expression in the following manner:

> "I believe that we are on the threshold of a global revolution: that the benefits of a global market place, combined with effective international institutions, will set humanity on a course of increasing prosperity through technological innovation and societal evolution that we can only dare to dream of."

> (cited in Sullivan, 1997, p. 111)

This is not a consensual but a contested vision of globalisation. Nevertheless, its central starting points — the amalgam of the market place and social progress — have framed the Organisation's education policy agendas. Such a logic we believe has run its course in the face of what may be interpreted as the legitimation crisis of globalisation. Many no longer believe in the benefits of the global market place. Many dissent from the credo that the global market place, with its endemic social and economic polarities, constitutes an appropriate point of departure for social organisation, or for education. At this juncture the vacuum of policy and ideas stands evident. The problem lies not so much with instrumental framing of education for employment or economic growth. It is rather to be found in the way in which market logic has been allowed so to capture the education policy agenda as to render education systems and actors virtually powerless to cultivate alternative frames of reference for education — and most particularly those which take critical intellectual inquiry and the fostering of an egalitarian and democratic ethos as their starting point. Without these perspectives more humane visions and versions of globalisation will not be realised. Given the struggle within the OECD over educational priorities and purposes, and given its long history as a think-tank and catalyst of ideas, it will surely be interesting to see how the Organisation, with its expanded membership and networks of influence, now deals with the dilemmas of globalisation when working towards a new set of educational priorities for the 21st century.

References

Alexander, T.J., 1994. Introductory address. In: CERI, *Making Education Count. Developing and Using International Indicators.* OECD, Paris, pp. 13–18.

Anderson, B., 1983. *Imagined Communities.* Verso, London.

APEC, 1997. *School Based Indicators of Effectiveness: Experiences and Practices in APEC Members (APEC Education Forum).* Published for APEC: Guangxi Normal University Press, Guilin, PR China.

Appadurai, A., 1996. *Modernity at Large: Cultural Dimensions of Globalisation.* University of Minnesota Press, Minneapolis.

Apple, M., 1992. Review of OECD (1989) Education and the economy in a changing society. *Comp. Educ. Rev.* 36(1): 127–129.

Archer, C., 1994. *Organizing Europe: The Institutions of Integration* (2nd ed.). Edward Arnold, London.

Ashton, D. and Green, F., 1996. *Education, Training and the Global Economy.* Edward Elgar, Cheltenham.

Atchoarena, D., 1995. Conclusions and implications for education policies. In: Atchoarena D. (Ed.), *Lifelong Education in Selected Industrialized Countries.* International Institute for Educational Planning (UNESCO), Paris, and National Institute for Education Research, Tokyo, pp. 216–231.

Atkinson, P., 1990. *The Ethnographic Imagination: Textual Constructions of Reality.* Routledge, London.

Australian Education Council Review Committee, 1991. *Young People's Participation in Post-Compulsory Education and Training.* (Finn Committee Report). AGPS, Canberra.

Australian Education Council/MOVEET, 1992. *Putting General Education to Work. The Key Competencies Report.* (Mayer Committee Report). AGPS, Canberra.

Axtmann, R. (Ed.), 1998a. *Globalization and Europe. Theoretical and Empirical Investigations.* Pinter, London.

Axtmann, R., 1998b. Globalisation, Europe and the State: Introductory Reflections. In: Axtmann, R. (Ed.), *Globalisation and Europe: Theoretical and Empirical Investigations.* Pinter, London, pp. 1–22.

Back, K., Davis, D. and Olsen, A., 1996. *Internationalisation and Higher Education: Goals and Strategies*. IDP Education Australia, Evaluations and Investigations Program, Higher Education Division, Department of Employment, Education and Youth Affairs, Canberra.

Ball, S., 1990. *Politics and Policy Making in Education: Explorations in Policy Sociology*. Routledge, London.

Ball, S., 1994. *Education Reform. A Critical and Post-structural Approach*. Open University Press, Buckingham.

Ball, S., 1996. Performativity and fragmentation in 'post-modern schooling'. In: Carter, J. (Ed.), *Post-modernity and the Fragmentation of Welfare: A Contemporary Social Policy*. Routledge, London.

Ball, S., 1997. Policy sociology and critical social research: A personal review of recent education policy and policy research. *Br. Educ. Res. J.* 23(3): 257–274.

Ball, S., 1998. Big policies/small world: An introduction to international perspectives in education policy. *Comp. Educ.* 34(2): 119–130.

Barnett, K., 1993. *Swings and Roundabouts. The Open Training Market and Women's Participation in TAFE*. Department of Employment, Education and Training, Canberra.

Barthes, R., 1967. *Writing Degree Zero*. Cape, London.

Bauer, M. and Kogan, M., 1997. Evaluation systems in the UK and Sweden: Successes and difficulties. *Eur. J. Educ.* 32(2): 129–143.

Bauman, Z., 1998. *Globalization: The Human Consequences*. Polity Press, Cambridge.

Beazley, Hon. K., 1992. *International Education in Australia through the 1990s*. AGPS, Canberra.

Beck, U., 1992. *Risk Society: Towards a New Modernity*. Sage, London.

Bernstein, B., 1990. *The Structure of Pedagogic Discourse (Class, Codes and Control, Vol IV)*. Routledge, London.

Bessis, S., 1995. *From Social Exclusion to Social Cohesion: A Policy Agenda*. Management of Social Transformations Policy Papers, No. 2. UNESCO, Paris.

Bhabha, H., 1994. *The Location of Culture*. Routledge, London.

Bienefeld, M., 1996. Is a strong national economy a utopian goal at the end of the twentieth century? In: Boyer, R. and Drachem, D. (Eds.), *States Against Markets. The Limits of Globalization*. Routledge, London, pp. 415–440.

Blackmore, J., 1999. *Troubling Women. Feminism, Leadership and Educational Change*. Open University Press, Buckingham.

Boci, J. and Thomas, G.M. (Eds.), 1999. *Constructing World Culture International Nongovernmental Organizations since 1875*. Stanford University Press, Stanford.

Bottani, A., 1995. Comparing educational output. *The OECD Observer* 193, April/May, 6–11.

Bottani, N. and Walberg, H.J., 1992. What are international indicators for? In: CERI, *The OECD International Education Indicators. A Framework for Analysis*. OECD, Paris, pp. 7–12.

Bowe, R., Ball, S. and Gold, A., 1992. *Reforming Education and Changing Schools: Case Studies in Policy Sociology*. Routledge, London.

Bowles, S. and Gintis, H., 1976. *Schooling in Capitalist America*. Routledge and Kegan Paul, London.

Boyer, E.L., 1990. *Scholarship Reconsidered: Priorities of the Professorate*. The Carnegie Foundation for the Advancement of Teaching, Princeton, NJ.

Braverman, H., 1974. *Labor and Monopoly Capital: The Degradation of Work in the Twentieth Century*. Monthly Review Press, New York.

Bremer and van der Wende, 1995. Not in refs.

Brine, J., 1999. *Undereducating Women: Globalizing Inequality*. Open University Press, Buckingham.

Brown, P. and Lauder, H., 1996. Education, globalization and economic development. *J. Educ. Policy* 11(1): 1–25.

Brown, P., Halsey, A.H., Lauder, N. and Stuart Wells, A., 1997. The transformation of education and society: An introduction. In: Halsey, A.H., Lauder, H., Brown, P. and Stuart Wells, A. (Eds.), *Education Culture Economy Society*. Oxford University Press, Oxford, pp. 1–44.

Callahan, R., 1962. *Education and the Cult of Efficiency: A Study of the Social Forces That Have Shaped the Administration of the Public Schools.* University of Chicago Press, Chicago.

Cannon, R.A., 1997. Advancing International Perspectives: The Internationalisation of Higher Education in Indonesia. In: Murray-Harvey, R. and Silins, H. (Eds.), *Learning and Teaching in Higher Education: Advancing International Perspectives.* Proceedings of the HERDSA conference, Adelaide, pp. 51–71.

Cantor, L., 1974. *Recurrent Education. Policy and Development in OECD Member Countries. United Kingdom.* OECD/CERI, Paris.

Carley, M., 1980. *Rational Techniques in Policy Analysis.* Gower, Aldershot.

Castells, M., 1996. *The Rise of the Network Society.* Blackwell, Oxford.

CERI, 1992. *Education at a Glance. The OECD Indicators.* OECD, Paris.

CERI, 1993. *Education at a Glance. The OECD Indicators.* OECD, Paris.

CERI, 1994a. *Quality in Teaching.* OECD, Paris.

CERI, 1994b. *Making Education Count. Developing and Using International Indicators.* OECD, Paris.

CERI, 1995a. *Schools Under Scrutiny.* OECD, Paris.

CERI, 1995b. *Education at a Glance. The OECD Indicators.* OECD, Paris.

CERI, 1996a. *Education at a Glance. The OECD Indicators.* OECD, Paris.

CERI, 1996b. *Education at a Glance. Analysis.* OECD, Paris.

CERI, 1998a. *Education at a Glance. The OECD Indicators* OECD, Paris.

CERI, 1998b. *Education Policy Analysis.* OECD, Paris.

CERI/OECD, 1973. *Recurrent Education: A Strategy for Lifelong Learning.* OECD, Paris.

Cerny, 1990. *The Changing Architecture of Politics: Structure, Agency and the Future of the State.* Sage, London.

Clarke, J. and Newman, J., 1997. *The Managerial State.* Sage, London.

Commonwealth Department of Education, 1976 *Report of the Working Party on the Transition from Secondary Education to Employment.* AGPS, Canberra.

Considine, M., 1988. The corporate management framework as administrative science: a critique. *Aust. J. Public Administration* 47(1): 4–19.

Cope, B. and Kalantzis, M., 1998. Facing Our Educational Futures. *Education Australia Magazine,* http://www.edoz.com.au/edoz/editorial/edit.html

Craft, A. (Ed.), 1994. *International Developments in Assuring Quality in Higher Education.* Falmer Press, London.

Cunningham, S., Tapsall, S., Ryan, Y., Stedman, L., Bagdon, K. and Flew, T., 1997. *New Media and Borderless Education: A Review of the Convergence Between Global Media Networks and Higher Education Provision.* Department of Employment, Education, Training and Youth Affairs, AGPS, Canberra.

Dale, R., 1997. The state and the governance of education: An analysis of the restructuring of the state–education relationship. In: Halsey, A.H., Lauder, H., Brown, P. and Stuart Wells, A. (Eds.), *Education Culture Economy Society.* Oxford University Press, Oxford, pp. 273–282.

Dale, R., 1999. Specifying globalization effects on national policy: A focus on the mechanisms. *J. Educ. Policy* 14(1): 1–18.

Davies, S. and Guppy, N., 1997. Globalization and educational reforms in Anglo-American democracies. *Comp. Educ. Rev.* 41(4): 435–459.

Deacon, B., Hulse, M. and Stubbs, P., 1997. *Global Social Policy. International Organizations and the Future of Welfare.* Sage, London.

De Tocqueville, 1835. Please complete.

Delors, J., 1996. *Learning: The Treasure Within: Report to UNESCO of the International Commission on Education for the Twenty-first Century.* UNESCO, Paris.

Department of Education and Youth Affairs, 1983. *Youth Policies, Programs and Issues.* AGPS, Canberra.

Dias, M., 1994. Comments from UNESCO. In: Craft, A. (Ed.), *International Developments in Assuring Quality in Higher Education.* Falmer Press, London, pp. 156–167.

Dill, D., 1998. Evaluating the 'evaluative state': Implications for research in higher education. *Eur. J. Educ.* 33(3): 361–377.

Drucker, P., 1993. *Post-capitalist Society*. Harper, New York.

Duke, C., 1974. *Recurrent Education: Policy and Development in OECD Member Countries — Australia*. OECD/CERI, Paris.

Employment and Skills Formation Council, National Board of Employment, Education and Training, 1992. *Australian Vocational Certificate Training System*. (Carmichael Report). AGPS, Canberra.

Falk, R., 1993. The making of global citizenship. In: Brecher, J., Childs, J.B. and Cutler, J. (Eds.), *Global Visions: Beyond the New World Order*. Black Rose Books, Montreal, pp. 39–50.

Finegold, D., McFarland, L. and Richardson, W. (Eds.), 1993. *Something Borrowed, Something Learned? The Transatlantic Market in Education and Training Reform*. The Brookings Institute, Washington, DC.

Foucault, M., 1991. Governmentality. In: Burchell, G., Gordon, C. and Miller, P. (Eds.), *The Foucault Effect: Studies in Governmentality*. Harvester Wheatsheaf, Hemel Hempstead, pp. 87–104.

Fowler, F., 1995. The international arena: The global village. *J. Educ. Policy* 9(5/6): 89–102.

Fraser, N., 1995. From redistribution to recognition: Dilemmas of justice in a 'post-socialist' society. *New Left Rev.* July–August, 68–93.

Friedman, T., 1999. *The Lexus and the Olive Tree*. Harper Collins, New York.

Garrett, G. and Mitchell, D., 1996. *Globalisation and the welfare state: Income transfers in the industrial democracies, 1966–90*. Paper presented to the Annual Meeting of the American Political Science Association, San Francisco, CA.

Gewirtz, S., Ball, S. and Bowe, R., 1995. *Markets, Choice and Equity in Education*. Open University Press, Buckingham.

Giddens, A., 1994. *Beyond Left and Right: The Future of Radical Politics*. Polity Press, Cambridge.

Giddens, A., 1998. *The Third Way. The Renewal of Social Democracy*. Polity Press, Cambridge.

Green, A., 1999. Education and globalization in Europe and East Asia: convergent and divergent trends. *J. Educ. Policy* 14(1): 55–72.

Green, D., 1994. Trends and issues. In: Craft, A. (Ed.), *International Developments in assuring Quality in Higher Education*. Falmer Press, London, pp. 168–177.

Haarlov, 1997. Not in ref. list.

Haas, E., 1990. *When Knowledge is Power. Three Models of Change in International Organizations*. University of California Press, Berkeley.

Habermas, J., 1996. The European Nation-State — its achievements and its limits. On the past and future of sovereignty and citizenship. In: Balakrishnan, G. (Ed.), *Mapping the Nation*. Verso, London, pp. 281–294.

Hadenius, A. and Uggla, F., 1998. Shaping civil society. In: Bernard, A., Helmich, H. and Lehning, P. (Eds.), *Civil Society and International Development*. OECD, Paris.

Halpin, D., 1994. Practice and prospects in education policy research. In: Halpin, D. and Troyna, B. (Eds.), *Researching Education Policy: Ethical and Methodological Issues*. Falmer Press, London, pp. 198–206.

Halsey, A.H., Lauder, H., Brown, P. and Stuart Wells, A. (Eds.), 1997. *Education Culture Economy Society*. Oxford University Press, Oxford.

Harding, S. (Ed.), 1994. *The 'Racial' Economy of Science*. Indiana University Press, Bloomington.

Harding, S., 1998. *Is Science Multicultural? Postcolonialisms, Feminisms and Epistemologies*. Indiana University Press, Bloomington.

Harvey, D., 1989. *The Condition of Postmodernity*. Blackwell, Oxford.

Henry, M. and Taylor, S., 1993. Gender equity and economic rationalism. In: Lingard, B., Knight, J. and Porter, P. (Eds.), *Schooling Reform in Hard Times*. Falmer Press, London, pp. 153–175.

Henry, M. and Taylor, S., 1995. Equity and the AVC pilots in Queensland: A study in policy retraction. *Aust. Educ. Res.* 22(1): 85–106.

Henry, M., Lingard, B., Rizvi, F. and Taylor, S., 1999. Working with/against globalisation in education. *J. Educ. Policy* 14(1): 85–97.

Heyneman, S.P., 1993. Quantity, quality and source. Presidential address. *Comp. Educ. Rev.* 37(4): 372–388.

Hirsch, D., 1996. Report of the Maastricht Conference. In: *Knowledge Bases for Education Policies*. OECD, OECD Documents, Paris, pp. 22–32.

Hirst, P. and Thompson, G., 1996. *Globalisation in Question*. Polity Press, Cambridge.

Hobsbawm, E., 1994. *Age of Extremes: The Short Twentieth Century 1914–1991*. Michael Joseph, London.

Holton, R., 1998. *Globalization and the Nation-State*. MacMillan Press, Basingstoke.

Hood, B., 1995. Emerging issues in public administration. *Public Administration* 73, Spring: 165–183.

IMHE, 1999. *Programme on Institutional Management in Higher Education Home Page*. http://www.oecd.org//els/edu/imhe/index.htm

Istance, D., 1996. Education at the Chateau de la Muette. *Oxford Rev. Educ.* 22(1): 91–96.

Johnson, R., 1976. Notes on the schooling of the English working class, 1780–1850. In: Dale, R., Esland, G. and MacDonald, M. (Eds.), *Schooling and Capitalism: A Sociological Reader*. Routledge and Kegan Paul, London, pp. 44–54.

Kangan, M. (chairperson), 1975. *TAFE in Australia: Report on needs in technical and further education*. AGPS, Canberra.

Kearns, P. and Hall, W. (Eds.), 1994. *Kangan: 20 years on*. National Centre for Vocational Education Research, South Australia.

Kemmis, S., Marginson, S., Porter, P. and Rizvi, F., 1999. *Enhancing diversity in Australian higher education: A discussion paper*. A paper prepared for the Senate External Strategy Committee, University of Western Australia.

Kickert, W., 1991. Steering at a distance: A new paradigm of public governance in Dutch higher education. Paper presented to European Consortium for Political Research, University of Essex.

Kishun, R., 1998. Internationalization in South Africa. In: Scott, P. (Ed.), *The Globalization of Higher Education*. The Society for Research into Higher Education and Open University Press, Buckingham, pp. 58–69.

Knight, J. and de Wit, H., 1995. Strategies for internationalisation of higher education: historical and conceptual perspectives. In: de Wit, H. (Ed.), *Strategies for Internationalisation of Higher Education*. EAIE, Amsterdam.

Knight, J. and Lingard, B., 1997. Ministerialisation and politicisation: Changing structures and practices of education policy. In: Lingard, B. and Porter, P. (Eds.), *A National Approach to Schooling in Australia*? Australian College of Education, Canberra, pp. 26–45.

Kogan, M., 1979. *Education Policies in Perspective: An Appraisal*. OECD, Paris.

Kogan, M., 1995. Policy lessons. *Times Educ. Suppl.*, March.

Latham, M., 1998. *Civilizing Global Capital. New Thinking For Australian Labor*. Allen and Unwin, Sydney.

Lauder, H., Hughes, D., Watson, S., Waslander, S., Thrupp, M., Strathdee, R., Simiyu, I., Dupuis, A., McGlinn, J. and Hamlin, J., 1999. *Trading in Futures. Why Markets in Education Don't Work*. Open University Press, Buckingham.

Lenn, M., 1994. International linkages and quality assurance: A shifting paradigm. In: Craft, A. (Ed.), *International Developments in Assuring Quality in Higher Education*. Falmer Press, London, pp. 127–133.

Levitas, R., 1996. The concept of social exclusion and the new Durkheimian hegemon. *Crit. Soc. Policy* 16(1): 5–20.

Levitas, R., 1998. *The Inclusive Society? Social Exclusion and New Labour*. MacMillan Press, Basingstoke.

Lingard, B., 1999 It is and it isn't: Vernacular globalisation, educational policy and restructuring. In: Burbules, N. and Torres, C. (Eds.), *Globalisation and Educational Policy*. Routledge, New York.

Lingard, B., Knight, J. and Porter, P. (Eds.), *Schooling Reform in Hard Times*. Falmer Press, London.

Lister, R., 1998. From equality to social inclusion: New Labour and the welfare state. *Crit. Soc. Policy* 55 18(2): 215–225.

Loya, T.A. and Boli, J., 1999. Standardization in the world policy: Technical rationality over power. In: Boli, J. and Thomas, G. (Eds.), *Constructing World Culture International Nongovernmental Organizations since 1975*. Stanford University Press, Stanford, pp. 169–197.

Lukes, S., 1983. *Power. A Radical View*. MacMillan Press, London.

Lyotard, J.F., 1984. *The Postmodern Condition: A Report on Knowledge*. Manchester University Press, Manchester.

Maasen, P., 1997. Quality in European higher education: Recent trends and their historical roots. *Eur. J. Educ.* 32(2): 111–127.

Marginson, S., 1997a. *Educating Australia. Government, Economy and Citizen since 1960*. Cambridge University Press, Cambridge.

Marginson, S., 1997b. *Markets in Education*. Allen and Unwin, Sydney.

Marginson, S., 1999. After globalization: Emerging politics of education. *J. Educ. Policy* 14(1): 19–32.

Markoff, J., 1998. Globalization and the future of democracy. Paper at 14th World Congress of Sociology, Montreal, Canada.

Martin, H. and Schumann, H., 1997. *The Global Trap*. Zed Books, London.

McCarthy, C., 1998. *The Uses of Culture Education and the Limits of Ethnic Affiliation*. Routledge, New York.

McCollow, J. and Lingard, B., 1996, Changing discourses and practices of academic work. *Aust. Univ. Rev.* 39(2): 11–19.

McKenzie, P., 1983. *Recurrent Education: Economic and Equity Issues in Australia*. ACER, Melbourne.

McNeely, C. and Cha, Y., 1994. Worldwide educational convergence through international organizations: Avenues for research. *Educ. Policy Anal. Arch.* 2(14).

Meadmore, D., 1998. Changing the culture: The governance of the Australian pre-millennial university. *Int. Stud. Sociol. Educ.* 8(1): 27–45.

Mestenhauser, J., 1997. On moving cemeteries and changing curricula. *Newslett. Eur. Assoc. Int. Educ.*

Meyer, J.W., Kamens, D., Benavot, A., Cha, Y. and Wong, S., 1992a. *School Knowledge for the Masses: World Models and National Primary Curricular Categories in the Twentieth Century*. Falmer Press, London.

Meyer, J.W., Ramirez, F.O. and Soysac, Y., 1992b. World expansion of mass education, 1870–1980. *Sociol. Educ.* 65(2): 128–149.

Meyer, J.W., Boli, J., Thomas, G. and Ramirez, F., 1997. World society and the Nation-State. *Am. J. Sociol.* 103(1): 144–181.

Miyoshi, M., 1998. 'Globalisation', culture, and the university. In: Jameson, F. and Miyoshi, M. (Eds.), *The Cultures of Globalization*. Duke University Press, Durham, pp. 247–270.

Muetzelfeldt, M., 1995. Democracy, citizenship and the problematics of governing production: The Australian case. In: Jureidini, R. (Ed.), *Labour, Unemployment and Democratic Rights*. Centre for Citizenship and Human Rights, Geelong, pp. 33–48.

Neave, G., 1988. On the cultivation of quality, efficiency and enterprise: An overview of recent trends in higher education in Western Europe, 1986–1988. *Eur. J. Educ.* 23(1/2): 7–23.

Neave, G., 1991. On programmes, universities and Jacobins: Or, 1992 vision and reality for European higher education. *Higher Educ. Policy* 4(4): 37–41.

Neave, 1993. Not in references. Please provide complete reference.

Neave, G., 1997. Global civilization and cultural values: Apocryphal or millennial visions? Introduction, *Global Civilization and Cultural Roots: Bridging the Gap — the Place of International University Cooperation, Report of the IAU Tenth General Conference 1995*. International Association of Universities, Paris.

Nuttall, D., 1992. The functions and limitations of are international education indicators. In: CERI, *The OECD International Education Indicators. A Framework for Analysis*. OECD, Paris, pp. 13–23.

Oakes, J., 1986. *Educational Indicators: A Guide for Policy Makers*. The Rand Corporation, Santa Monica, CA.

OECD, 1975. *Education, Inequality and Life Chances*. OECD, Paris.

OECD, 1976. *Educational Policy and Planning: Transition from School to Work or Further Study in Australia*. OECD, Paris.

OECD, 1977. *Australia. Transition from School to Work or Further Study*. OECD, Paris.

OECD, 1979. *Future Educational Policies in the Changing Social and Economic Context*. OECD, Paris.

OECD, 1981. *Activities of OECD in 1980. Report by the Secretary-General*. OECD, Paris.

OECD, 1982. *Activities of OECD in 1981. Report by the Secretary-General*. OECD, Paris.

OECD, 1983a. *Activities of OECD in 1982. Report by the Secretary-General*. OECD, Paris.

OECD, 1983b. *Policies for Higher Education in the 1980s*. OECD, Paris.

OECD, 1984. *Activities of OECD in 1983. Report by the Secretary-General*. OECD, Paris.

OECD, 1985a. *OECD*. OECD, Paris.

OECD, 1985b. *Activities of OECD in 1984. Report by the Secretary-General*. OECD, Paris.

OECD, 1985c. *Education in Modern Society*. OECD, Paris.

OECD, 1986a. *Girls and Women in Education*. OECD, Paris.

OECD, 1986b. *Youth and Work in Australia*. OECD, Paris.

OECD, 1987a. *Activities of OECD in 1986. Report by the Secretary-General*. OECD, Paris.

OECD, 1987b. *Innovation in Education No. 45*. OECD, Paris.

OECD, 1987c. *Innovation in Education No. 47*. OECD, Paris.

OECD, 1987d. *Universities Under Scrutiny*. OECD, Paris.

OECD, 1987e. *Structural Adjustment and Economic Performance*. OECD, Paris.

OECD, 1989a. *Draft Programmes of Work for 1990, Education Committee, Governing Board of the Centre for Educational Research and Innovation*. OECD, Paris.

OECD, 1989b. *Education and the Economy in a Changing Society*. OECD, Paris.

OECD, 1989c. *Education and Structural Change: A statement by the Education Committee*. OECD, Paris.

OECD, 1989d. *Schools and Quality*. OECD, Paris.

OECD, 1990a. *Financing Higher Education: Current Patterns*. OECD, Paris.

OECD, 1990b. *Labour Market Policies for the 1990s*. OECD, Paris.

OECD, 1991. *Medium-Term Priorities and Draft Programmes of Work for 1992, Directorate for Social Affairs, Manpower and Education, Centre for Educational Research and Innovation*. OECD, Paris.

OECD, 1992a. *Activities of OECD in 1991. Report by the Secretary-General*. OECD, Paris.

OECD, 1992b. *Innovation in Education No. 62*. OECD, Paris.

OECD, 1992c. *Innovation in Education No. 63*. OECD, Paris.

OECD, 1992d. *Draft Programmes of Work for 1993, Education Committee, Governing Board of the Centre for Educational Research and Innovation*. OECD, Paris.

OECD, 1992e. *High-Quality Education and Training For All*. OECD, Paris.

OECD, 1993a. *Activities of OECD in 1992. Report by the Secretary-General*. OECD, Paris.

OECD, 1993b. *Draft Programmes of Work for 1994, Education Committee and Governing Board of the Centre for Educational Research and Innovation*. OECD, Paris.

OECD, 1993c. *OECD Contribution: Background Report*. The Transition from Elite to Mass Higher Education Conference, Sydney, 15–18 June.

OECD, 1994a. *The OECD*. OECD, Paris.

OECD, 1994b. *Draft Programmes of Work for 1995/1996 Education Committee and Governing Board of the Centre for Educational Research and Innovation*. Directorate for Education, Employment, Labour and Social Affairs, OECD, Paris.

OECD, 1994c. *Women and Structural Change.* OECD, Paris.

OECD, 1994d. *School: A Matter of Choice.* OECD, Paris.

OECD, 1994e. *Draft Programme of Work for 1995, Programme on Institutional Management in Higher Education Directing Group.* OECD, Paris.

OECD, 1994f. *Progress Report, Centre for Educational Research and Innovation.* OECD, Paris.

OECD, 1994g. *Vocational Education and Training for Youth: Towards Coherent Policy and Practice.* OECD, Paris.

OECD, 1995a. *Activities of OECD in 1994. Report by the Secretary-General.* OECD, Paris.

OECD, 1995b. *Performance Standards in Education. In Search of Quality.* OECD, Paris.

OECD, 1995c. *Governance in Transition: Public Management Reforms in OECD Countries.* OECD, Paris.

OECD, 1995d. *Education At a Glance III in the Press.* OECD, Press Review Section, Paris.

OECD, 1996a. *Activities of OECD in 1995. Report by the Secretary-General* OECD, Paris.

OECD, 1996b. *Draft Programmes of Work for 1997–1998, Education Committee and Governing Board of CERI.* Directorate for Education, Employment, Labour and Social Affairs, OECD, Paris.

OECD, 1996c. *OECD Work of Education, Employment, Labour and Social Affairs.* OECD, Paris.

OECD, 1996d. *Lifelong Learning For All.* OECD, Paris.

OECD, 1996e. *Indicateurs Internationaux Des Systems D'Enseignment: Propositions En Vue De La Prochaine Phase 1997–2001, Comite de L'Education, Comite Directeur Du CERI.* OECD, Paris.

OECD, 1996f. *Globalisation and Linkages to 2030: Challenges and Opportunities for OECD Countries.* OECD, Paris.

OECD, 1996g. OECD indicators chart progress towards inclusive and effective education systems, OECD News Release, Paris, 27 Nov. 1996, http://www.oecdwash.org/PRESS/PRERELS/pub9683.htm

OECD, 1997a. *Education at a Glance: OECD Indicators.* OECD, Paris.

OECD, 1997b. *Education and Equity in OECD Countries.* OECD, Paris.

OECD, 1997c. *Towards a New Global Age: Challenges and Opportunities.* OECD, Paris.

OECD, 1997d. *Societal Cohesion and the Globalising Economy: What Does the Future Hold.* OECD, Paris.

OECD, 1997e. *Thematic Review of the Transition from Initial Education to Working Life: Australia, Country Note.* http://www.oecd.org/copyr.htm

OECD, 1997f. *Ministerial Council Communique.* OECD, Paris.

OECD, 1998a. *Annual Report 1997.* OECD, Paris.

OECD, 1998b. *Education Catalogue.* OECD, Paris.

OECD, 1998c. *Redefining Tertiary Education.* OECD, Paris.

OECD, 1998d. *Civil Society and International Development.* OECD, Paris.

OECD, 1998e. *Pathways and Participation in Vocational and Technical Education and Training.* OECD, Paris.

OECD, 1999. Not in refs.

OECD, undated. *The OECD.* OECD, Paris.

OECD undated, circa 1997. *OECD.* OECD, Paris.

OECD/CERI, 1983. *Quality in Education.* OECD, Paris.

OECD/CERI, 1985. *Integration of the Handicapped in Secondary Schools: Five Case Studies.* OECD, Paris.

OECD/CERI, 1987. *Multicultural Education.* OECD, Paris.

OECD/CERI, 1988. *Disabled Youth: The Right to Adult Status.* OECD, Paris.

OECD/CERI, 1989. *One School, Many Cultures.* OECD, Paris.

OECD/CERI, 1991. *Disabled Youth: From School to Work.* OECD, Paris.

OECD/CERI, 1994. *Education in a new internationalisation: Curriculum Development for Internationalisation (Project Brief).* OECD, Paris.

OECD/CERI, 1995a. The INES Project (1988–1995) (Historical Brief). Unpublished paper.

OECD/CERI, 1995b. The lifelong learner in the 1990s. In: Atchoarena, D. (Ed.), *Lifelong Education in Selected Industrialized Countries*. International Institute for Educational Planning (UNESCO), Paris, and National Institute for Education Research, Tokyo, pp. 201–215.

OECD/CERI, 1996. *Internationalisation of Higher Education*. OECD, Paris.

Ohmae, K., 1995. *The End of the Nation State: the Rise of Regional Economies*. Free Press, New York.

Ozga, J., 1987. Studying education through the lives of policy makers: An attempt to close the micro–macro gap. In: Walker, S. and Barton, L. (Eds.), *Changing Policies: Changing Teachers*. Open University Press, Milton Keynes, pp. 138–150.

Pair, C., 1998. Synthesis of Country Reports. In: OECD *Pathways and Participation in Vocational and Technical Education and Training*. OECD, Paris, pp. 9–25.

Papadopoulos, G., 1994. *Education 1960–1990. The OECD Perspective*. OECD, Paris.

Peters, M. and Marshall, J., 1996. *Individualism and Community: Education and Social Policy in the Postmodern Condition*. Falmer Press, London.

Pocock, B., 1992. *Women in Entry Level Training. Some overseas experiences*. AGPS, Canberra.

Popkewitz, T., 1996. Rethinking decentralisation and state/civil society distinctions: The State as a problematic of governing. *J. Educ. Policy* 11(1): 27–51.

Prunty, J., 1984. *A Critical Reformulation of Educational Policy Analysis*. Deakin University Press, Geelong.

Pusey, M., 1991. *Economic Rationalism in Canberra: A Nation-building State Changes its Mind*. Cambridge University Press, Cambridge.

Putman, R., 1993. *Making Democracy Work. Civic Traditions in Modern Italy*. Princeton University Press, NJ.

Raffe, D., 1998a. Does learning begin at home? The use of 'home international' comparisons in UK policy making. *J. Educ. Policy* 13(5): 591–602.

Raffe, D., 1998b. Conclusion: Where are pathways going? Conceptual and methodological lessons from the pathways study. In: OECD *Pathways and Participation in Vocational And Technical Education and Training*. OECD, Paris, pp. 375–391.

Rawls, J., 1972. *A Theory of Justice*. Clarendon Press, Oxford.

Reich, R.B., 1991. *The Work of Nations: Preparing Ourselves for 21st Century Capitalism*. A.A. Knopf, New York, NY.

Rhodes, R.A.W., 1997. *Understanding Governance Policy Networks: Governance, Reflexivity and Accountability*. Open University Press, Buckingham.

Rizvi, F., 1994. Devolution in education: Three contrasting perspectives. In: Martin, R., McCollow, J., McFarlane, L., McMurdo, G., Graham, J. and Hull, R. (Eds.), *Devolution, Decentralisation and Recentralisation: The Structure of Australian Schooling*. Australian Education Union, Melbourne, pp. 1–5.

Robertson, R., 1992. *Globalization*. Sage, London.

Roche, G. and Marginson, S. (Eds.), 1979. *Preliminary Responses to the Williams Report*. Melbourne University Students' Representative Council, Melbourne.

Rose, N., 1990. *Governing the Soul: The Shaping of the Private Self*. Routledge, London.

Rose, N., 1996. The death of the social? Re-figuring the territory of government. *Econ. Soc.* 25(3): 327–356.

Rose, N., 1999. *Powers of Freedom: Reframing Political Thought*. Cambridge University Press, Cambridge.

Rose, N. and Miller, T., 1992. Political power beyond the state: Problematics of government. *Br. J. Sociol.* 43(2): 172–205.

Ruby, A., 1992. Do common values produce common indicators? In: CERI, *The OECD International Education Indicators. A Framework for Analysis*. OECD, Paris, pp. 77–82.

Ruby, A., 1997. How Was Paris and Three Other Questions About the OECD's Education Program, Symposium: The OECD, Globalization and Education Policy Making — The Case of Australia, Brisbane, 3–4 June.

Sachs, J., 1994. Strange yet compatible bedfellows: Quality assurance and quality improvement. *Aust. Univ. Rev.* 37(4): 22–25.

Schools Commission, 1973. *Schools in Australia, Report of the Interim Committee of the Australian Schools Commission.* Karmel Report. AGPS, Canberra.

Schugurensky, D., 1999. Higher education restructuring in the era of globalization: Towards a heteronomous model? In: Arnove, R.F. and Torres, C.A. (Eds.), *Comparative Education the Dialectic of the Global and the Local.* Rowman and Littlefield, New York, pp. 283–304.

Scott, P. (Ed.), 1998. *The Globalization of Higher Education.* Open University Press, Milton Keynes.

Seddon, T., 1994. *Context and Beyond: Reframing the Theory and Practice of Education.* Falmer, London.

Selden, R.W., 1992. Standardizing data in decentralized educational data systems. In: CERI, *The OECD International Education Indicators. A Framework for Analysis.* OECD, Paris, pp. 107–113.

Skilbeck, M., 1993. Opening Address. In: *The Transition from Elite to Mass Higher Education: An International Conference sponsored by the Australian Department of Employment, Education and Training in association with the Organisation for Economic Cooperation and Development.* Proceedings, AGPS, Canberra, pp. 17–22.

Sklair, L., 1996. Conceptualising and researching the transnational capitalist class in Australia. *Aust. N.Z. J. Sociol.* 32(2): 1–19.

Smyth, J. and Shacklock, G., 1998. *Re-Making Teaching. Ideology, Policy and Practice.* Routledge, London.

Spring, J., 1998. *Education and the Rise of the Global Economy.* Lawrence Erlbaum Associates, Mahwah, NJ.

Steedman, H., 1999. Measuring the quality of educational outputs: Some unresolved problems. In: Alexander, R., Broadfoot, P. and Phillips, D. (Eds.), *Learning from Comparing. New Directions in Comparative Education Research.* Symposium Books, Oxford, pp. 201–216.

Sullivan, S., 1997. *From War to Wealth. 50 years of innovation.* OECD, Paris.

Tanner, L., 1999. *Open Australia.* Pluto Press, Annandale.

Taylor, S. and Henry, M., 1994. Equity and the new post-compulsory education and training policies in Australia: A progressive or regressive agenda? *J. Educ. Policy* 9(2): 105–127.

Taylor, S. and Henry, M., 1996. Reframing equity in the Training Reform Agenda: implications for social change. *Aust. Vocat. Educ. Rev.* 3(2): 46–55.

Taylor, S., Rizvi, F., Lingard, B. and Henry, M., 1997. *Educational Policy and the Politics of Change.* Routledge, London.

Thomas, G.M., Meyer, J.W., Ramirez, F.O. and Boli, J., 1987. *Institutional Structure Constituting State, Society, and the Individual.* Sage, Beverly Hills.

Thomson, P., 1999. Towards a just future: Schools working in partnership with neighbourhoods made poor. Paper presented to Fifty UNESCO ACEID Conference *Reforming Learning, Curriculum and Pedagogy: Innovative Visions for the Next Century*, Bangkok, Thailand, 13–16 December.

Torres, C.A., 1995. State and education revisited: Why educational researchers should think politically about education. In: Apple, M. (Ed.), *Rev. Res. Educ.* 21, Washington: American Educational Research Association, pp. 255–331.

Townshend, J., 1996. An overview of OECD work on teachers, their pay and conditions, teaching quality and the continuing professional development of teachers. OECD paper presented at the UNESCO International Conference on Education, Geneva, 1996, unpaged.

Trow, 1974. Not in references. Please provide complete reference.

UNESCO, 1972. *Learning To Be: The Faure Report.* UNESCO, Paris.

UNESCO, 1994. *Policy Paper for Change and Development in Higher Education.* UNESCO, Paris.

UNESCO, 1996. *Medium-Term Strategy 1996–2001.* UNESCO, Paris.

UNESCO, 1998. Lifelong learning and training: A bridge to the future. Second International Congress on Technical and Vocational Education.

Urry, J., 1998. Contemporary transformations of time and space. In: Scott, P. (Ed.), *The Globalization of Higher Education*. The Society for Research into Higher Education and Open University Press, Buckingham, pp. 1–17.

Van Vught, F., 1994. Western Europe and North America. In: Craft, A. (Ed.), *International Developments in assuring Quality in Higher Education*. Falmer Press, London, pp. 3–17.

Vickers, M., 1994. Cross-national exchange, the OECD, and Australian education policy. *Knowledge Policy* 7(1): 25–47.

Vidovich, L. and Porter, P., 1997. The recontextualisation of 'quality' in Australian higher education. *J. Educ. Policy* 12(4): 233–252.

Walby, S., 1997. *Gender Transformations*. Routledge, London.

Waters, M., 1995. *Globalization*. Routledge, London.

Weiss, C., 1989. Congressional committees as users of analysis. *J. Policy Anal. Manage.* 8(3): 411–431.

Weiss, L., 1997. Globalisation and the myth of the powerless state. *New Left Rev.* 225: 3–27.

Weiss, M. and Weishaupt, H., 1999. The German School after Reunification: A descriptive overview of recent trends. *Int. J. Educ. Reform.* 8(2): 113–119.

Whitty, G., Power, S. and Halpin, D., 1998. *Devolution and Choice in Education: The School, the State and the Market*. ACER Press, Melbourne.

Williams, B., 1979. *Education, Training and Employment*. Report of the Committee of Inquiry into Education and Training. AGPS, Canberra.

Wiseman, J., 1998. *Global Nation? Australia and the Politics of Globalisation*. Cambridge University Press, Cambridge.

World Bank, 1991. *Vocational and Technical Education and Training: A World Bank Policy Paper*. World Bank, Washington.

Yeatman, A., 1990. *Bureaucrats, Technocrats, Femocrats: Essays on the Contemporary Australian State*. Allen and Unwin, Sydney.

Yeatman, A., 1994. *Postmodern Revisionings of the Political*. Routledge, New York.

Yeatman, A., 1998. Trends and opportunities in the public sector: A critical assessment. *Aust. J. Public Admin.* 57(4): 138–147.

Young, M.F.D., 1998. *The Curriculum of the Future. From the 'new sociology of education' to a critical theory of learning*. Falmer Press, London.

Subject Index